A History *of the* Universe *in* 21 Stars
(and 3 imposters)

A HISTORY *of the* UNIVERSE *in* 21 STARS
(and 3 *imposters)*

GILES SPARROW

WELBECK

Published by Welbeck
An imprint of Welbeck Non-Fiction Limited,
part of Welbeck Publishing Group.
20 Mortimer Street,
London W1T 3JW

First published by Welbeck in 2020

A CIP catalogue record for this book is available from the British Library

ISBN
Hardback – 9781787394650

Typeset by EnvyDesign
Illustrations by Laura Barnes, (L.B. Illustration), www.lbillustration.co.uk
Milky Way and long Exposure image © Shutterstock.com
Carina image © N. Smith / NOAO/AURA/NSF

Printed and bound by CPI Group (UK) Ltd., Croydon, CR0 4YY

10 9 8 7 6 5 4 3 2

www.welbeckpublishing.com

To Katja,
for support in
strange times

Contents

INTRODUCTION

Twinkle, twinkle little star...

*T*his, in case you hadn't guessed from the cover, is a book about stars.

They say that on a clear night, if you're well away from the modern bane of light pollution, enjoy a perfectly flat horizon and have been eating your carrots, you can perhaps see as many as 4,500 of them. The sky is full of the things, and on those rare occasions that you find yourself somewhere *really* dark on a truly clear night, there can be so many that you may find yourself floundering to get your bearings even if you're pretty familiar with the bright constellations (which is sort of my job).

Pick up a half-decent pair of binoculars, and the number of stars in your sky will instantly leap to more than a hundred thousand. A small telescope raises that number to 2.5 million plus – enough to keep the most obsessive stargazer busy for several lifetimes. But even these are just the tip of the cosmic iceberg – best estimates suggest that the Milky Way galaxy (the vast stellar disc we call home), contains perhaps 400 billion stars in all, with six or seven more sparking into life each year. And then you can probably square that figure of 400 billion, since there are at least as many galaxies in the Universe as there are stars in the Milky Way.

All of this suggests that stars are not an optional extra – those pretty lights in the night sky aren't just there for decoration. No, in fact we live on their goodwill – they're pretty much the only things in the Universe capable of producing heat and light to warm a planet's surface against the harsh indifferent cold of deep space. Plus, smaller worlds themselves wouldn't exist if it weren't for the gravitational driving force of star formation. Heat and light from space, along with geological energy from inside planets, are the only ways we know of powering the complex mess of biochemical reactions called life.

But our intimate relationship with the stars goes further than nurture – as Carl Sagan memorably put it, "we are made of star stuff". The book you're reading is made of atoms that probably passed several times through these great cosmic recyclers, as is the air you're breathing, the chair you're sitting in, and every molecule in your body (barring the hydrogen you inherited directly from the Big Bang itself).

Stars, in other words, are everything – so surely only an idiot would set out to tell the history of an estimated 160,000,000,000,000,000,000,000 of them through the lens of a mere 21? Fortunately, however, there are a couple of things tilting the scales in my favour.

First off, stars obey the laws of physics, just as surely as this book will if you drop it on your toe. Although each star is a unique case study, they all go through similar phases in a cycle of life and death, shine by the same basic processes, and tend to group together in distinct categories – all of which mean that what's true for one star will be more or less true for billions of others.

Second, while generations of stargazers have spent centuries putting the pieces of this story together, the pace has noticeably quickened since my days as an eager young astronomy graduate. Satellite observatories and computer-controlled giant telescopes have triggered an astronomical revolution – since the 1990s, we've been able to map the aftermath of the Big Bang, got to grips with the processes behind the birth and death of stars, discovered thousands of alien worlds, and found a whole new way of observing the distant cosmos through gravity rather than light*. As a result, I'm extraordinarily grateful that this book can draw on a vast pool of knowledge, theory and informed speculation.

Oh, and third, I've cheated. My limited handful of stars are seasoned with a light scattering of other objects: imposters that have all, at some point in their histories, been mistaken for stars. They're here now to help tell our story on the broadest possible canvas – the present, past and future of the Universe itself.

★ ★ ★

The stars we'll be visiting in the following pages were picked for a variety of reasons. Some, like 61 Cygni and Sirius B,

---

* Oh, and we've also discovered that something is causing space itself to stretch apart at an ever-increasing rate, but no biggie.

have played a unique role in the discovery of our place in the Universe. Others, such as Aldebaran and Eta Aquilae, are good representatives of broad families of objects, and help to tell the wider story. In most cases, however, it's a mix of the two.

Above all, though, I've tried to ensure that as many of these objects as possible are within easy reach. To see most of them, you'll need nothing more than a clear, dark sky and perhaps a phone app to help point the way (or the illustrations in this book). A handful of others can be spotted with a basic pair of binoculars or a small telescope. Only a couple, due to their very nature, are limited to the realm of the more serious amateur or pro astronomer.

Astronomy is both the oldest of the sciences, and the most engaging, for one very good reason: its accessibility. Any one of us can go out tonight and experience light rays from a distant star, ending a journey that may have begun thousands of years ago by striking the back of our retina and triggering a spark in our optic nerve. The vast scale of space, and the relative insignificance of our place within it, can be daunting, but it can also inspire a desire to ask questions and know more: *How I wonder what you are?* In these strange times of isolation, gazing at the skies can also offer the balm of a communal experience – in looking up at the same stars, we can find something to share with others near and far. So do get out there if you possibly can, and see how many of these 21 stars (and three imposters) you can spot for yourself.

Giles Sparrow, May 2020

1 – POLARIS

Learning the basics from the
laziest star in the sky

*L*et's start with an easy one.

Polaris, the northern pole star, is probably the most famous star in the sky, even if it's not actually the brightest. It also has the advantage that, if you're in the Northern Hemisphere, you should be able to spot it on any night of the year. If, on the other hand, you're south of the Equator, this is the one star in the book where you're sure to be out of luck – but hang in there and we'll get back to you shortly…

There are various ways of finding the northern pole star. If you want to be lazy you can just fire up a compass app on your phone and look for a moderately bright star on a line between due north on your horizon and the zenith (the point in the sky directly overhead).

But if your phone's out of charge or you simply like to do things the old-fashioned way, the traditional route is to use a brighter and more familiar group of stars to help guide you. The pattern of seven stars known as the Plough or Big Dipper is a permanent fixture in the sky across most of the Northern Hemisphere, swinging low over the northern horizon on autumn and winter evenings and hanging high overhead in summer. It isn't actually an official constellation in its own right, just the brightest part of the sprawling constellation Ursa Major, or the Great Bear.

Three of these seven stars form a curved handle, while the other four make a lopsided rectangle (either the blade of the plough, the bowl of the dipper, or a pan used to scoop up water, depending on your preference). The pair of stars furthest from the handle – Merak at the bottom and Dubhe at the top, assuming you're looking at the pattern the "right way up" – are known as "the pointers". Follow an imaginary line past Dubhe for about five times the Merak–Dubhe distance, and you'll come to a somewhat fainter star – that's Polaris.

Once you've done this a couple of times, it becomes second nature and you can soon dispense with the pointers entirely

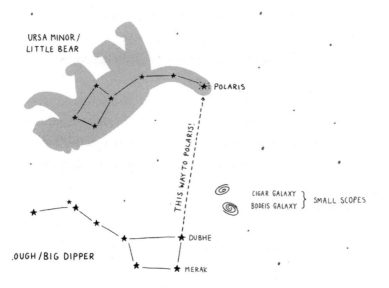

URSA MINOR / LITTLE BEAR

POLARIS

THIS WAY TO POLARIS!

CIGAR GALAXY
BODEIS GALAXY
} SMALL SCOPES

.OUGH / BIG DIPPER

DUBHE

MERAK

in favour of the northern pole star's own constellation, Ursa Minor or the Little Bear. As you might guess from the name, this looks quite like a smaller and fainter version of the Plough or Big Dipper, with a "tail" of three stars leading to a rectangle of four more. Conveniently enough, Polaris is the endmost star on the tail, and also the Little Bear's brightest star. Officially it's designated Alpha Ursae Minoris, following a scheme (invented by Germany's Johann Bayer for his 1603 star atlas Uranometria) that tags a constellation's brightest stars with sequential letters of the Greek alphabet.

Polaris stands out amongst all the stars in the heavens because it's a fixed point in the sky – the one star that barely moves. This is because it lies almost directly above Earth's own North Pole. If you could look at Earth from outside and draw a line through both poles, it would point to a spot in the sky very close to Polaris – the North Celestial Pole or NCP.

The NCP stays still because most of the movement of the stars and other objects, including the Sun and planets, has nothing to do with these objects themselves – it's almost entirely down to Earth's own rotation and movement through space. Earth is spinning on its axis (taking 23 hours and 56 minutes to make a complete rotation), but from your perspective, it seems like the sky is rotating in the opposite direction*. Stare upwards for even a few minutes, and you'll soon start to notice the stars drifting slowly from east to west, as your own location on Earth rolls inexorably eastwards.

Long-exposure photos of the sky demonstrate this beautifully, revealing the trails of stars as bright arcs across the

---

* Meanwhile the Sun drifts in the other direction to the tune of about four minutes, which is why we have a 24-hour day.

sky. Most stars appear from beneath the eastern horizon, reach their highest point in the sky as they cross a north-south line across the sky known as the meridian, and set in the west. But stars close enough to the celestial pole are "circumpolar" – they neither rise nor set, instead following circular tracks around the sky. For northern stargazers, the pole star marks the bullseye of these concentric rings, but the same effect applies equally in both hemispheres.

Just how high Polaris sits in the sky, and which stars and constellations are circumpolar, depends on latitude. This is your position on Earth's surface, measured in degrees north or south of the Equator. If you were at the North Pole itself (latitude 90°N), then the NCP would be directly overhead and all the stars in the sky would be circumpolar, following circular tracks parallel to the horizon without rising or setting. As you

move southwards, however, Polaris and the NCP slip gradually down the sky towards the northern horizon and the circle of circumpolar stars gets smaller*.

Now, seems as good a place as any to discuss angles in the sky. They're measured in just the same way as angles on Earth, if you remember your school geometry, with 360 degrees all the way around the sky†, and 90 degrees in a right angle (for instance between your horizon and the zenith point directly overhead). Each degree is further divided into 60 minutes of arc, and each minute into 60 seconds of arc (so you might have an angle of, say, 5° 32' 15" – this stands for for five degrees, 32 minutes, 15 seconds).

Hold out your arm as far as possible and spread your fingers, and that's *about* 10 degrees (roughly the width across the Big Dipper's "pan"). Clench your fist, and that's *about* five degrees (more or less the distance between Dubhe and Merak). Stick up your thumb, and it'll be *roughly* one degree wide. The Sun and a full Moon both have an average diameter of half a degree, and the limit of resolution (allowing you to distinguish between details) for good human eyesight is about one minute of arc.

Polaris sits roughly half a degree from the celestial pole itself, and as a result describes a very tight circle around the NCP. Considering this is a chance alignment with a star hundreds of trillions of kilometres away, we're lucky to have such a bright marker for the sky's central axis.

---

* A handy hint: whichever hemisphere you're in, your celestial pole sits above the horizon at an angle equal to your latitude.

† The system goes back around 4,000 years to the Mesopotamians, who liked everything to be in multiples of 60 because it's "multi-factorial", which is math-speak for "you can divide it up in a lot of different ways and still end up with a whole number". In the days before the Casio FX-80 calculator, this was a big deal, and made it easy to do a lot of sums in your head (or at least on a clay tablet).

In search of the southern pole star

The sky around the South Celestial Pole (SCP) is rather different from its northern equivalent, jumbled with faint stars in some fairly obscure constellations invented by French astronomer Nicolas-Louis de Lacaille during a mid-eighteenth-century stint at the Cape of Good Hope. The SCP itself lies in the constellation Octans, known as the Octant (an obsolete navigational instrument). A faint star called Sigma Octantis is the closest to the pole that's visible to the naked eye, but it's more than one degree away. Fortunately, there are a couple of other ways to find the sky's southern pole.

Follow the Southern Cross: The classic technique to find the southern pole star is to first identify the famous compact

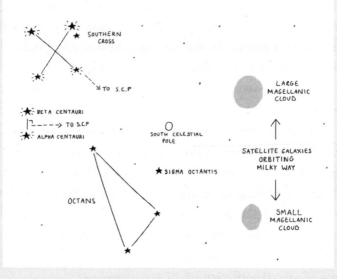

group of Crux Australis, the Southern Cross (beware of imitations as there are a couple of cross-like patterns to mislead the unwary). Draw an imaginary line along the long axis of the cross from Gacrux at the top to the brightest star Acrux at the bottom, and then extend it by about four and half times (you'll miss the pole by a few degrees, but still be in the right area).

Draw a bright star triangle: Find the two stars Canopus (the second brightest star in the entire sky, in the constellation of Carina) and Achernar (the bright star at the end of the river constellation Eridanus). Imagine an equilateral triangle with these stars at two of the corners, extending into the far southern sky. The SCP sits at the "missing" corner.

In the last few centuries BCE, Greek stargazers transformed the system of angular measurements in the sky into a theoretical model of the Universe where the Earth was surrounded by a series of concentric, nested spheres carrying the Sun, Moon, planets and stars.

The idea of these spheres began in the fourth century BCE with that great Hellenic brainbox Plato. Always a fan of the tricky question, he pondered whether the apparently unpredictable motions of the Sun, Moon and planets could in fact be explained by a set of interacting cycles, each of which involved circular motion at a uniform rate (for Greeks, ideas of circularity and uniformity not only made sums easier, they also fitted in with ideas of natural perfection).

Plato's idea seemed so appealing that his disciples spent the next few centuries coming up with attempts to make it work, invoking systems of nested crystal spheres and adding more and more sophistication in the hope that they might eventually find a model that correctly predicted the motions of the planets. The stars, at least, were easy – all they required was a single sphere, fixed at the celestial poles and rotating once each day.

In the second century CE, Ptolemy of Alexandria, a Greek-Egyptian polymath who we'll run into a few times in the course of our story, immortalised his own take on this Heath-Robinson Universe in the great astronomical textbook best known as the *Almagest*. A classical bestseller, it remained the last word on astronomy for almost 1,500 years, until a bunch of uppity Renaissance scholars dared to question first the Earth's position at the centre of the Universe, and then the sacrosanct principle of uniform circular motion.

By the early 1600s, thanks largely to the work of German astrologer Johannes Kepler[†], elliptical paths or "orbits" around the Sun were the in thing. With every object now capable of changing its distance from the Sun, there was no place for the planetary spheres, but the concept was so convenient that astronomers stuck with the idea of an outermost sphere of "fixed stars" surrounding the Earth. To this day, this celestial sphere provides a co-ordinate system against which everything else can be measured.

★ ★ ★

---

[*] A title applied to it by later Arab astronomers, meaning "The Great Work". Not bad as reviews go, and certainly catchier than the original *Mathematical Syntaxis*.

[†] Prior to the eighteenth century, astrology and astronomy were pretty much interchangeable, since almost everyone researching the stars was doing so for purposes of prediction – Galileo was a notable exception.

The simplest way of thinking about the celestial sphere is as an extension of Earth's familiar co-ordinate systems onto an imaginary spherical shell that wraps the entire sky. In reality, the distance to stars, planets and other objects varies wildly, but for our point of view on Earth, all we're concerned with is their direction, so we can simplify things by imagining them moving on this shell. Celestial poles (above the geographical north and south poles) mark the pivot points, while midway between them runs a celestial Equator, splitting the sky, like the Earth, into the Northern and Southern Hemispheres.

Stargazers on Earth get to see different parts of the celestial sphere depending on time and location. It goes without saying that half of the sphere is blocked at any one time by the stuff you're standing on, but looking towards the appropriate celestial pole, you can see circumpolar stars spinning around a fixed point in the sky, while in the opposite direction you can watch stars rise from the eastern horizon, cross over the meridian and set towards the west. The celestial equator passes from due east to due west and crosses the meridian at an angle linked to our particular latitude*. Beneath it, stargazers away from the poles get to see stars in the opposite celestial hemisphere.

The orientation of the sky, and the stars you see at a certain time of night, also change through the year because our system of timekeeping is based on the Sun rather than the stars. As Earth orbits around it each year, the Sun slowly changes direction against the more distant stars, moving westwards through the band of constellations known as the zodiac so that anything close to it is lost in the glare. Stars and planets first emerge from

---

* The Equator's maximum altitude in your local sky is simply 90° minus your own latitude.

their close encounters with the Sun into the eastern morning sky, then track slowly westwards over the months as their separation increases, before eventually sinking into the evening sunset as the Sun once again closes in on their position.

The Sun's track around the sky is called the ecliptic – though in reality, it is the plane of Earth's own orbit around the Sun. And because Earth's poles are tipped over at an angle relative to this plane, the ecliptic in the sky is tilted at 23.5° to the celestial equator. Therefore, the Sun spends half the year in the sky's Northern Hemisphere and half in the Southern Hemisphere, giving each of Earth's hemispheres longer days in turn, and crossing over at the intersection points called equinoxes.

But the vagaries of Earth's orbit mean that Polaris wasn't always the pole star, and it won't be again in future. That planetary tilt, which tips the poles at 23.5 to the ecliptic, slowly changes direction as the gravity of the Sun and Moon tug at the 20-kilometre bulge around the Equator.* As a result, the north and south poles follow a lazy circle know as axial precession, lasting 25,772 years, and the celestial poles wander around the sky in the same period. Polaris happens to be in the firing line of the NCP at the moment, but 4,000 years ago, Kochab, the Little Bear's second-brightest star, was closer. Around 12,000 years from now, a really bright star called Vega, in the constellation of Lyra, will come within fourdegrees of the pole. Southern stargazers, meanwhile, only have to wait another 5,000 years before their celestial pole passes close to three bright stars in fairly rapid succession.

★ ★ ★

---

* Like Asterix's best friend Obelix, Earth's not fat, but its chest has slipped a bit – due to our planet's fast spin, the Equator is literally trying to fly away into space.

So what about Polaris itself? Is it just a dull star that happened to be in the right place at the right time? Happily, that's not the case – and in fact the northern pole star is a great example of a couple of different types of object we'll be encountering in more detail in later chapters. For one thing, Polaris is a variable star – it doesn't shine with steady brightness but instead pulsates slightly, brightening and then fading in a cycle of around four days.

Star brightness is measured using a system called magnitude – a scale on which lower magnitudes are brighter than higher ones. In ancient times, the brightest stars of all were pegged as first magnitude, while the faintest visible to the naked eye were of sixth magnitude˙. By this reckoning, Polaris is a star of mid-third magnitude, but fortunately these days we can be a bit more precise than that.

In 1856, a young astronomer called Norman Pogson, who had run away from a career in the Nottingham lace trade to become a scientist, worked out that there was a hundred-fold difference in brightness between a typical first-magnitude and a sixth-magnitude star. He formalised the system with a precise factor of 2.512 between each magnitude division (because $2.512^5 = 100$), and calibrated the whole thing by assigning Polaris a magnitude of precisely $2.0^{1\dagger}$. This meant that the brightest stars in the sky, such as Sirius and Canopus, suddenly gained *negative* magnitudes, since they were so much brighter than Polaris.

---

* We've inherited this system from Ptolemy, so blame him, or perhaps the earlier Greek Hipparchus (second century CE) who is often attributed with its invention (although anything he had to say on the subject has long since been lost to posterity).

† Later, when the northern pole star's brightness turned out to be a bit wobbly, astronomers switched their calibration point to the rather more reliable Vega as magnitude 0.0.

In this modern magnitude scale (called *apparent* magnitude since it measures the appearance of stars from Earth), Polaris actually wobbles between magnitude 1.86 and 2.13, averaging out at 1.98. Like many stars, its changes are due to pulsations, but while most pulsating stars are red, Polaris is yellow. In fact, it's an example of a type of object called a Cepheid variable, about which we'll learn a lot more when we visit the star Eta Aquilae.

Another thing worth noting is that Polaris, like many bright stars, is not alone. While nearly all the light we see comes from the main star (officially Polaris Aa), it has two much smaller companions in space, each a little hotter and brighter than our Sun. One of these companions (Polaris B) was discovered in 1779 and can be spotted through a decent telescope, while the second (Polaris Ab) is too close to Polaris Aa to be seen with anything short of the Hubble Space Telescope[2]. From this distance, Polaris B and Polaris Ab shine at magnitudes 8.7 and 9.2 respectively – below naked-eye visibility.

The latest measurements of the northern pole star's distance (achieved using a technique called parallax, which lies at the heart of our next chapter) suggest it is some 447 light years from Earth – so far away that the photons of light sparking your optic nerve when you look at Polaris set out on its journey to Earth when Elizabeth I was on the throne of England[3]. Even though professional astronomers can be a little sniffy about it, the light year is a handy way of describing vast astronomical distances, and we'll be sticking with it throughout this book. In more everyday terms, it's about 9.5 million million km – the distance that light, the fastest thing in the Universe, travels in an average year.

Since we know how bright the stars look from Earth, that means we can work out how bright it *really* is. Polaris Aa turns out to average about 2,500 times brighter than the Sun*, putting it in a class of stars known as supergiants.

However, the northern pole star has one final mystery up its sleeve. Because it's been observed for so long and so meticulously, astronomers can look at records as far back as Ptolemy himself, and track how it's changed over time. And this suggests that Polaris's average brightness has increased considerably – perhaps by as much as two and a half times (a whole order of magnitude) over the past couple of thousand years. More accurate and recent measurements, meanwhile, suggest that Polaris's pulsations have been getting generally smaller as it has brightened (they almost stopped in the 1990s, but have since increased again)[4].

Such changes would be very unusual – aside from regular pulsations, stars just aren't supposed to go through this sort of major shift in brightness over what is, in astronomical terms, a relatively short period of time. Assuming the changes are real, then perhaps by chance we're catching Polaris on the cusp of a significant threshold in its evolution, as a sea change in the internal processes of energy generation makes itself felt in the star's overall energy output. We'll look at more of these key moments in a star's life when we come to the famous pulsating star Mira in a later chapter.

"But I am constant as the northern star, of whose true-fix'd and resting quality, there is no fellow in the firmament," says Julius Caesar in Shakespeare's play. Wrong on both counts, it would seem.

---

* Sounds impressive? Wait until we get to Eta Carinae…

2 – 61 Cygni

Measuring the distance to a
flying star

---

*I*f Earth is really spinning around the Sun, then why can't we feel it? That question puzzled scholars and philosophers from the time of Ancient Greece, and became increasingly pressing when a dying Polish priest called Nicolaus Copernicus released his new theory of the Universe into the wild in 1543.

This isn't the place to go into a blow-by-blow account of the Copernican Revolution, but one of many common-sense Renaissance objections to the idea of a Sun-centred (rather than Earth-centred) model of the Universe was a perfectly reasonable question: why doesn't our changing point of view affect the directions of the stars through the year. The answer is that it does – but only very slightly. Conclusive evidence arrived very late to the party, long after the Copernican debate had been settled by the formidable tag team of Galileo Galilei and Johannes Kepler*,

---

*   Galileo famously spotted moons around Jupiter, phases on Venus, and other new discoveries that undermined the Ptolemaic view. More or less simultaneously, Kepler realised that if planetary orbits were ellipses rather than the perfect circles suggested by Copernicus, then the Sun-centred system could actually be used for practical predictions.

but the quest to detect the shift remained a major project for astronomers over almost two centuries. It was seen as a door through which the scale of the Universe itself could be revealed, and an inconspicuous star called 61 Cygni would prove to be the key.

Compared to romantic-sounding names like Polaris, Rigel and Aldebaran, 61 Cygni seems dull. That's because, despite being a naked-eye star at magnitude 5.2, it is easily overlooked and was not catalogued until astronomer Royal John Flamsteed set out to do things methodically in the late seventeenth century from the swanky new Royal Observatory at Greenwich. Wisely recognising that no one would want to learn a whole new set of star names and designations from scratch, he opted instead to "plug the gaps" left where the great star mapper Johann Bayer ran out of Greek letters or simply couldn't be bothered chasing down the faintest stars. Flamsteed methodically catalogued the "left behind" stars by their right ascension (the coordinate by which the position of a star is measured), tracking from west to east across each constellation in turn. This required him to draw up strict lines and divide the constellations into areas of the sky rather than subjective patterns, but it produced a coherent system that still survives today.

As you might guess from its relatively high number, 61 Cygni lies in a constellation – Cygnus, or the Swan – packed with naked-eye stars. This large and prominent pattern does roughly resemble a long-necked bird flying southwards down the Milky Way, with the bright star Deneb marking its tail feathers in the north and Albireo (a beautiful orange-and-blue double star) its beak to the south. It's a familiar constellation of northern summer and autumn, passing almost directly overhead, while Southern

Hemisphere stargazers get to see it sail over their northern horizon on evenings between August and October.

61 Cygni sits just behind the Swan's outstretched western wing. The best way to find it is to look along the line from Sadr, the central star of Cygnus's cross-shape, towards epsilon a little further southeast. Tucked behind this line is a neat little right-angled triangle made up of zeta, nu and tau. 61 Cyg lies just past halfway along the line linking nu to tau.

At magnitude 4.8, you should be able to spot our star with the naked eye under dark skies once your eyes have adapted to the dark, but binoculars will show it more easily under city lights. If they've got a power of 10x or more and you can keep your hands steady, they should also reveal the star's first secret – it's a double consisting of two orangey-coloured stars, one a bit brighter at magnitude 5.2, the other somewhat fainter at magnitude 6.1.

61 Cygni's double nature was spotted for the first time by astronomer James Bradley in September 1753. Over the following decades, the stars were occasionally visited by astronomers keen to investigate the nature of such close pairs, but it was not until 1792 that Italian Giuseppe Piazzi, a Catholic priest and astronomer whose recently founded Palermo Observatory was equipped with state-of-the-art equipment for measuring stellar positions, noticed something strange: the twin stars had shifted position, and now lay slightly but unmistakeably northeast of the location reported by Bradley[1].

At the time, Piazzi made a note of this unusual drift through the sky (a phenomenon that astronomers call proper motion*),

---

* As opposed to the illusory motion caused by Earth's rotation and orbit around the Sun.

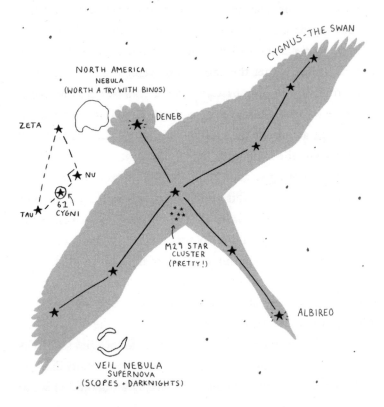

NORTH AMERICA
NEBULA
(WORTH A TRY WITH BINOS)

CYGNUS-THE SWAN

ZETA

DENEB

NU

61
CYGNI

TAU

M29 STAR
CLUSTER
(PRETTY!)

ALBIREO

VEIL NEBULA
SUPERNOVA
(SCOPES + DARKNIGHTS)

but he did not confirm it until 1804, when he revisited Cygnus during his compilation of a detailed stellar catalogue. In the intervening years, Piazzi had found fame through his discovery of Ceres, the largest asteroid and the first object to be found orbiting between Mars and Jupiter. Careful checking now confirmed his suspicion that 61 Cygni was moving across the sky at a surprisingly rapid rate of 4.1 seconds of arc per year (equivalent to the width of an average full Moon every 464 years). After the catalogue was published in 1806, 61 Cygni soon garnered the nickname of "Piazzi's Flying Star".

Based on the reasonable assumption that stars travel at similar speeds through space, you can use proper motion as a neat proxy for a star's likely distance from Earth – the closer the star is, the bigger the angle its movement across the sky is likely to cover in a year. 61 Cygni turned out to have the largest proper motion yet discovered, and was therefore immediately recognized as one of the closest stars to Earth. This made it a perfect target for ongoing attempts to measure our shifting cosmic point of view.

The posh word for this annual shift is parallax. For a quick demo, raise one finger of an outstretched hand and wink with one eye and then the other to see how its apparent direction shifts against more distant objects in the background. Now imagine your finger is somewhere out beyond Alpha Centauri, while your eyes are straddling Earth's orbit and separated by 300 million kilometres. The distance to even the nearest stars is about 135,000 times bigger than the "baseline" across Earth's orbit, so you can probably imagine how small the shift in angle would be in a situation like that – less than one second of arc or 1/3600th of a degree*.

These minute angles explain why astronomers had so much trouble spotting them, and their elusive nature was periodically used as a stick with which to beat the Copernican system, even after most sensible people had become converts. The obvious

---

* Astronomers actually use parallax as the basis of their preferred distance measuring system: one parsec is the distance (equivalent to 3.26 light years) at which an object shows a parallax of exactly one". Parsecs are handy for pros because taking the reciprocal of an object's parallax in arc seconds (1/the parallax) gives you its distance in parsecs without any more tedious maths: ½" parallax = two parsecs distance and so on. But light years are so ingrained in how most of us think about the Universe that, in this book, we're sticking with them.

solution was to accept that the stars were much further away than anyone had thought – incredibly far beyond the orbit of the most distant planets. However, proving the existence of parallax remained a sore point, and a hobby horse for many talented astronomers.

Among the challenges these parallax hunters faced were the relatively primitive quality and low power of their telescopes, the effects of the atmosphere blurring the sharpness of star images, and even the simple difficulty of knowing where your telescope was actually pointing[*].

Early attempts at finding a solution therefore employed some ingenious lateral thinking. James Bradley (long before his encounter with 61 Cygni) figured out a way of testing the parallax of a moderately bright star called Eltanin or Gamma Draconis[2]. His approach measured the angle between Eltanin and the zenith (the point directly overhead) with pinpoint accuracy at the exact moment the star crossed the north-south line across the sky, so he could be certain of Eltanin's precise position.

Bradley chose Eltanin because it passes almost directly overhead from London, helping to minimise another trouble-some effect we haven't even mentioned yet – atmospheric refraction. As if the parallax quest didn't throw up enough challenges, astronomers also have to deal with the fact that, as well as rippling and blurring the light from stars, Earth's atmosphere also deflects light rays onto new paths. We've all seen refraction at the boundary between air and water, for instance when groping for the last teaspoon in the washing-up, but it also affects starlight entering Earth's atmosphere from space.

---

[*] This was in the days before accurate telescope mounts and mechanisms existed to turn a telescope in sync with Earth's rotation.

The refraction effect is strongly linked to a star's altitude in the sky, because when we look at stars near the horizon we're looking through a much thicker depth of atmosphere than if we look straight up. There's a neat equation to model this, but Earth's atmosphere is notoriously changeable so for precision measurements, it's better to simply avoid the effect as far as possible by looking at objects almost directly overhead.

When Bradley and his collaborator Samuel Molyneux began measuring Eltanin's position in December 1725, they soon found that the star was moving. But from the outset their results were puzzling – Eltanin was moving southwards at a time when parallax should have already put it at its southernmost point in the sky. By March, its motion finally slowed and reversed, and it then tracked northwards until September when it reversed again. Two more years of observation confirmed the turning points were consistently three months of step, with Eltanin always switching direction in March and September rather than June and December as predicted.

At first, Bradley wondered if they had discovered a small annual wobble in the direction of Earth's poles, which would make the stars seem to inscribe small circles or ovals on the sky. But as he studied the evidence further, he realised they had in fact unearthed an entirely unconnected piece of evidence for Earth's motion – an effect known as aberration of starlight. This is a change in the angle at which starlight approaches Earth due to our motion around the Sun (if you think about raindrops falling at a steady angle while you stand still, and then imagine how their angle changes when you're walking in different directions, you'll get the picture). Because Earth's axis in space points in a constant direction as we make our annual trip around the Sun, the angle at which starlight falls onto Earth changes slightly in spring and autumn[3].

The complexities of aberration added further burdens to the already tough task of measuring parallax, but astronomers are a persistent breed, and thus the search continued throughout the eighteenth century and into the nineteenth, occasionally enlivened by triumphant announcements, nitpicking rebuttals and embarrassed retractions. Piazzi himself fell victim to one such mistake (almost certainly due to refraction) when he thought he'd measured the parallax of Sirius at four seconds of arc in 1808[4]*.

It was another three decades, however, before technology and skilful observing finally delivered an irrefutable parallax measurement. In the end, the race came down to two of the nineteenth century's finest observers – Friedrich von Struve and Friedrich Wilhelm Bessel. Both benefited from the fact that by this time, clockwork-driven telescope mounts had been invented that could keep pace with the sky's apparent rotation and stop their target stars rapidly drifting out of the field of view of a high-magnification eyepiece. They both also followed the suggestion, made by William Herschel in the 1780s,[†] that the best approach was to look for changes in a target star's position relative to others that lay nearby in the sky, rather than trying to keep track of its precise location on the celestial sphere.

---

* A figure that, if correct, would have put it less than 10 light months from Earth – about a factor of 10 out.

† William Herschel is chiefly known today as the discoverer of Uranus, but as we'll see elsewhere, his influence stretches far beyond being unwitting godfather to a thousand bad jokes.

Struve's effort focused on Vega, among the brightest and best known of all stars, while Bessel concentrated on the far more obscure 61 Cygni. As it turns out, Vega is about twice as far away as 61 Cygni, and has slightly less than half the parallax, making Struve's task considerably more difficult.

Struve was also using a traditional "micrometer" eyepiece – a design with a heritage dating back to the 1640s, in which two fine parallel wires are projected into the observer's field of view. The separation between these wires can be changed very gradually by turning an adjustment screw, and you can then translate the distance between the wires into an angular separation in the sky with some fairly simple maths. Using this apparatus, Struve began to track Vega's relative motion in late 1835. By 1837, he had 17 measurements that allowed him to publish a preliminary figure for Vega's parallax of one-eighth of a second of arc – very close to the modern value. Had he stopped there he might perhaps have claimed the prize, but instead he continued and by 1840, when he published his final result, his estimate had doubled, putting it far adrift of later measurements.

Bessel, in contrast, used a different sort of setup known as a heliometer. This was a refracting (lens-based) telescope whose main or objective lens was carefully cut into two halves. The separation produced a double image in the eyepiece, and one half-lens could be finely adjusted with a screw system, so that when the images of two separate stars lined up, the heliometer revealed the angular separation between them.

Beginning in August 1837, Bessel managed to take 98 parallax measurements of 61 Cygni over just 13 months. Wasting no time, he quickly processed his data and published his results by way of a letter to Sir John Herschel, President of the Royal Astronomical Society in London, on 23 October 1838.[5]*

Bessel's calculations were a tour de force, and immediately convincing in a way that Struve's so-far limited data failed to be. He not only estimated the parallax of the overall 61 Cygni system as 0.314" (equivalent to 10.3 light years), but also analysed the relative motions of the two stars and showed that they took at least 540 years to orbit each other. These figures stand up remarkably well even today, where the system's parallax has been refined to 0.286", its distance to 11.4 light years and its orbital period to around 678 years.

John Herschel referred to Bessel's measurements as the moment when the "sounding line in the universe of stars had at last touched bottom." – They marked the beginning of a new era in which stars were transformed from points of light in the sky into distant but measurable objects whose physical properties could be analysed and understood. For instance, now their distance from Earth was known, the intrinsic brightness of 61 Cygni's near-twin stars could be calculated. The brighter star proved to be less than one-sixth of the brightness of the Sun and the fainter less than one-tenth as bright, undermining older speculations that the difference in the brightness of stars might purely be down to their distances. In modern terms, 61 Cygni A and B are both

---

* In a curious case of synchronicity, a *third* astronomer was also hot on the parallax trail in 1838. Scotsman Thomas Henderson (1798–1844) had actually taken the necessary measurements of the bright Southern Hemisphere star Alpha Centauri while working at the Cape of Good Hope in the early 1830s. Using a zenith instrument very similar to that devised by Bradley, he had by 1833 successfully detected the star's annual north-south parallax drift. However, mindful of the many previous false alarms, he held off publishing until more complete measurements could confirm the motion in right ascension, and didn't get his result out until 1839.

orange dwarf stars (we'll look in more detail at what this means when we come to Proxima Centauri, the closest star of all).

As to the quest for parallax, it would be nice to say that the trickle of measurements in the late 1830s opened the floodgates for a torrent of others, but the reality was rather different. Parallax calculations remained demanding and elusive for all but the nearest stars until well into the twentieth century. Perhaps just 20 in total were known by the 1880s, and a further 180 in the decades that preceded World War I. Even at that point, astronomer Royal Frank W. Dyson estimated measurable parallax was limited to around 0.02", putting anything more than 160 light years from the Sun beyond the reach of direct measurement.

Until the Space Age, parallax was only able to provide a foundation to stellar astronomy – a key to the distance and physical properties of a very limited number of stars. Fortunately, the patterns of characteristics this relative handful of stars revealed were sufficient for the rough distance of many others to be reverse-engineered (for more on this, see Alcyone).

The dawn of space-based observing, of course, has proved too great an opportunity for parallax-hungry astronomers to miss. Atelescope positioned outside of Earth's atmosphere is able to make measurements with incredible precision, ignoring the challenges of refraction and atmospheric turbulence to deliver pin-sharp measurements down to a limit determined only by its size. The first dedicated parallax satellite, Hipparcos, was launched by the European Space Agency in 1989 and operated until 1993, delivering high-precision data for 118,000 stars and less accurate figures for 2.4 million more. Since 2013, Hipparcos has been succeeded by Gaia, an even more ambitious mission aiming to catalogue the distance of one billion objects all the way to the centre of our galaxy (26,000 light years away) and beyond.

Parallax remains our only method of directly measuring the distance to objects in the wider cosmos, and provides a happily secure first rung on a ladder of cosmic distances that gets increasingly rickety as it extends further from the certainties of Earth. However, as we'll see, it's still our best hope for understanding the complexities of the Universe as a whole, so we should be thankful for what certainty we have – and spare a thought for the obscure double star in Cygnus where it all began.

3 – ALDEBARAN

*How the colour of a giant reveals
its hidden secrets*

⸺✦⸺

*A*longside brightness, the most obvious outward sign that stars are physically different from each other is colour – stand outside on a clear night, and it won't take you long to spot a few variations. Some are blatantly obvious, but many others have more subtle distinctions – binoculars will help, and if you can get two contrasting stars in the same field of view that's a great way of bringing out the colour difference.

Aldebaran is one of the brightest stars in the sky, located in one of the sky's most recognisable star patterns, but its orange colour is what makes it really stand out. Embedded like a flaming beacon in the midst of a v-shaped cluster of stars called the Hyades, it marks the wild eye of the charging bull Taurus – a constellation recognised by stargazers for perhaps 18,000 years or more.

The star's name comes from the Arabic *al Dabarān*, meaning "the follower" – perhaps because it seems to follow the Pleiades (the famous star cluster that we'll come to when we visit Alcyone, Aldebaran's near neighbour), across the sky. We'll be coming back to Taurus on several occasions throughout this book, so

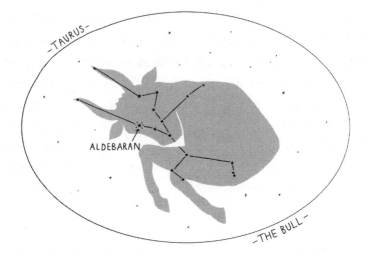

we'll save the mythology for later. For now, it's enough to say that Aldebaran begins to become visible in eastern pre-dawn skies around July and slowly tracks westwards, away from the Sun and into evening skies, where it can be seen from around November to April.

Look up Aldebaran in an online star catalogue, and you'll see it described as a K5+ III star. K5+ is its "spectral class" – a rough description of that fiery orange colour. The "III", meanwhile, is Aldebaran's "luminosity class" – technically speaking, this star is a giant.

All this terminology might sound a bit like the astronomical equivalent of stamp collecting – do we really *need* to catalogue and classify stars with obscure letters and numbers? Well, if we want to get to grips with the relationship between different types of stars and understand the story of how they live and evolve, I'm afraid we do, but I'll try to be as gentle as possible. In this chapter we'll be concentrating mostly on the story of *spectral* classes – we'll come back to luminosity classes in Alcyone.

The present-day jargon arose from the discovery, in the latter half of the nineteenth century, that starlight contains far more information than you might think. If astronomy up to this point was concerned with measuring the positions and movements of stars through space, then the arrival of new ways to analyse starlight and discover the physical properties of stars themselves marked the beginning of its sister discipline: astrophysics. And Aldebaran's bright colour and vivid hue meant it was there right at the start of this new branch of science.

* * *

The story begins in 1666, the year that the Great Plague ravaged Britain. As cities shut down and people fled to the countryside, a 24-year-old Isaac Newton exiled himself from Cambridge to his mum's place in Lincolnshire. However, young Isaac had more on his mind than getting his laundry done, and he was soon busy revolutionising physics.

As if messing about with apples and discovering gravity wasn't enough, Newton also found the time to investigate the properties of light, passing sunbeams through a prism and inspiring the cover art for *The Dark Side of the Moon*. What he was actually doing was splitting (and then reuniting) a beam of white sunlight to show that it was composed of different colours.

The prism split the light thanks to an optical effect called refraction, which bends colours towards the violet end of the spectrum more than those towards the red end. Figuring out the cause of these different colours, however, took a century and a half, much of it occupied with arguments about whether light itself was a bullet-like corpuscle, as Newton claimed, or a moving disturbance in space called a wave. Doubts lingered until 1821, when French civil engineer Jean-Augustin Fresnel topped out his finest work – a comprehensive wave theory of light.

Fresnel explained colour as a property that depends on the wavelength of light our eyes are receiving. Think about light as a moving ripple like a water wave, and the wavelength is the distance between two successive peaks or troughs: longer wavelengths are perceived as reddish colours, while shorter wavelengths skew to the blue. It's worth noting, though, that *all* these wavelengths are incredibly small by everyday standards, varying between about 400 and 700 *billionths* of a metre.

A few years before Fresnel's theoretical breakthrough, meanwhile, an inventive German glassmaker called Joseph von Fraunhofer* had made a practical leap that would prove to be

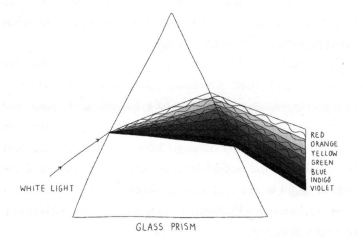

RED
ORANGE
YELLOW
GREEN
BLUE
INDIGO
VIOLET

WHITE LIGHT

GLASS PRISM

---

* Fraunhofer lived a pretty extraordinary life: orphaned at 11, he was apprenticed to a bullying glassmaker, then pulled from the rubble of a collapsed factory by the Prince-Elector of Bavaria, who saw the lad's potential and supported his education. Young Joseph proved to be a prodigy, and his breakthroughs led to huge improvements in the precision of optical instruments, but he died early, falling victim to the toxic fumes associated with his profession.

just as important. Fraunhofer wanted to study the spectrum in detail, and realised it wasn't enough to just stick a prism into a broad beam of light as Newton had done – the dispersed colours from different parts of the beam overlapped each other and washed out detail. So, he narrowed down the beam by passing it through the thinnest possible slit, then sent it through a prism made from glass of his own secret recipe. He then studied the emerging beam through a telescope-like eyepiece that could pivot to see different parts of the refracted beam.

When Fraunhofer used this device (which we would today call a spectroscope) to study sunlight, he discovered that its rainbow-like "continuum" of colourful light was crossed by hundreds of narrow dark lines, varying in strength and intensity. This was weird – it suggested that the Sun either wasn't producing certain very specific colours of light, or that these colours were somehow being prevented from reaching Earth.

The Fraunhofer lines remained a curiosity for more than 40 years, and their eventual explanation fell neither to astronomers nor opticians, but to a chemist and a physicist. Robert Bunsen was the chemist – a giant of a man from the German state of Lower Saxony whose daredevil investigations had made him the world's leading expert on arsenic (and cost him the sight in one eye). Gustav Kirchhoff, the physicist, was a dapper, humorous Prussian, well known already for his studies of electrical circuits.

Bunsen and Kirchhoff were both interested in the properties of chemical elements, and especially the light they gave out when heated. They used Bunsen's recently invented gas burner to heat highly refined samples in a clean, fierce flame, and soon found that each substance produced light with a few very specific colours that appeared through a spectroscope as bright "emission lines" on a dark background – a unique barcode in the chemical supermarket.

Around 1859, Kirchhoff and Bunsen tried shining bright white light through the heated vapour. What they saw was eerily familiar: a rainbow-like continuum criss-crossed by dark lines whose positions matched the bright lines of the material's emission spectrum. The material in the flame was clearly soaking up certain wavelengths of light, creating an "absorption spectrum". From there, it was easy for the duo to show that Fraunhofer's dark lines were created through the absorption of light by various familiar elements. For the first time, the chemical make-up of the Sun had, at least in part, been revealed.[*]

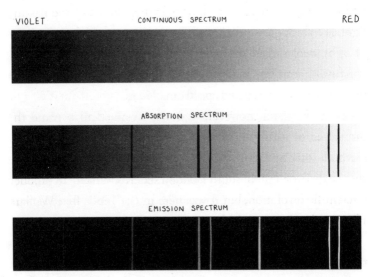

* At this point in proceedings, spare a thought for poor old Auguste Comte (1798–1857), the first real "philosopher of science" who, as late as 1835, had suggested that both the chemical composition and the surface temperatures of stars would be forever unknowable. Presumably he hadn't heard about Fraunhofer's lines.

Bunsen and Kirchhoff's work caused something of a sensation among astronomers, and many leapt to put this new tool to use. Italy's Giovanni Battista Donati was quick off the mark – attaching a spectroscope to his telescope in 1860, he soon began spotting prominent dark lines in the spectra of fifteen stars. Aldebaran and nearby Betelgeuse in Orion were among his targets, chosen for their brightness and distinctive reddish colour.

New Yorker Lewis Rutherfurd followed up in 1862 with an approach that would soon be widely adopted. Instead of using a prism to spread the light, he used a device called a diffraction grating – a blackened glass plate scratched with a series of fine lines to create parallel transparent windows. The grating used a different principle from the prism to give a similar result, and had two big advantages: it didn't absorb as much of the passing light as a thick glass prism, and the angle over which the colours of light were spread depended only on the spacing of the tiny windows. Pack them in tightly enough, and you could get a much broader and more detailed spectrum.

It was England, however, that was destined to become the birthplace of astrophysics, and to point the way towards new methods of classifying the stars. The unremarkable Victorian suburb of Tulse Hill in south London seemed unlikely to produce a scientific revolution, but it was here in the 1860s that William Huggins lived and worked.

The well-to-do son of a city silk merchant, Huggins had (in that typically Victorian self-starting way) determined in his late twenties that he would make A Contribution to Science. After dabbling in microscopy and physiology, he eventually settled on astronomy, building a small observatory in the back garden of the family home. After several years of methodical but unremarkable observations, the new breakthroughs in spectroscopy provided him with just the project he needed.

Working with his neighbour, chemistry professor William Allen Miller, Huggins set out to discover the composition of the stars. The project faced huge challenges – not just the faintness of starlight when spread through a spectroscope, but also the fact that no two spectroscopes dispersed light in exactly the same way. There was not yet any way of comparing lines recorded in the laboratory with those seen in starlight, so Huggins and Miller constructed an ingenious instrument that allowed them to burn chemical samples at the top of their telescope. Each observing session began by igniting a sample of sodium in the apparatus to create a "live" emission spectrum, and noting the position of its brightest emission lines as seen through the spectroscope. In this way, they could use lines of a known position to calibrate the absorption spectra of the stars they went on to observe.

Although Huggins and Miller measured spectra for almost 50 stars, their initial paper of 1864 chose to concentrate on just a handful. Aldebaran (perhaps just by virtue of its position in the alphabet) therefore became the first star to have its chemical barcode properly scanned.[1]

The pair identified 70 lines in the white, yellow, orange and red parts of the star's spectrum. There were clearly many more towards the blue end, but here the background light grew so faint that they were impossible to pin down. The measurable lines matched up to emissions from nine different elements – sodium, magnesium, hydrogen, calcium, iron, bismuth, tellurium, antimony and mercury. Comparisons with the spectra of seven other elements, meanwhile, seemed to rule out their presence.

The discovery that Aldebaran contained elements just like those on Earth and in the Sun finally put to rest any lingering doubts that the stars were suns like our own – but it would be another 60 years before the true balance of elements was entirely understood, and a century before their real significance was established.

That didn't necessarily stop astronomers from putting them to work in various ways, of course. For instance, Huggins, basking in his newfound reputation, soon announced another stunning feat, claiming to have found a tiny deviation in one of the hydrogen lines of Sirius from its proper wavelength. This, he said, was evidence for a phenomenon predicted by Austrian scientist Christian Doppler a few decades before – a slight change to the wavelength of starlight reaching Earth, created when the star is moving towards or away from us.

Spectral lines offered an ideal way of measuring this shift, since their precise "stationary" wavelengths could be pinned down from samples on Earth. We'll have much more to say about this when we come to Mizar, but for now it's enough to record that Huggins is usually credited as the first person to use the Doppler effect to measure a star's motion (even though we now know his figures for Sirius were out by a factor of four, and moving in the wrong direction).

Other astronomers, meanwhile, concentrated on simply cataloguing as many stellar spectra as possible. Over a decade from 1863, for instance, Jesuit priest Angelo Secchi mapped the spectra of 4,000 stars from a rooftop observatory on top of the Church of Saint Ignatius, a baroque pile in the heart of Rome[2]. Secchi soon identified three separate classes of stars, distinguished by the colour, intensity and distribution of their spectral lines. In this scheme, Aldebaran was an archetypal "Class II" star – a member of the same group that included the Sun, with yellow or orange colours and many lines linked to the presence of metals. Secchi's groupings weren't exactly nuanced, since any classification that groups a star like Aldebaran together with one like the Sun is missing out on some pretty big differences (as we'll see later, when we visit our own star), but it was a start. Just a few years after its heyday, however, Secchi's work would be dwarfed thanks to a technological revolution.

* * *

Astronomers had been playing with the idea of photographing the heavens ever since Louis Daguerre captured the first image of the Moon in 1839, but light-sensitive chemicals had remained messy and the entire process too slow and cumbersome to be regarded as anything other than a curiosity. That all changed in the 1870s with the invention of the "dry plate" process.

This technique for setting the necessary gunk in a gelatin film allowed plates to be mass-produced and stored before use, but also led to a massive improvement in their sensitivity*. Stargazers gleefully embraced the new technology – for the first time, it was not only possible to keep a permanent and precise record of the heavens, but also to soak up more light than the eye could see, revealing stars beyond the limits of human vision.

Spectroscopy, forever struggling with the challenges of spreading faint starlight out into fainter spectra, benefited most from the photography boom. One of its early converts was Henry Draper, a Virginia-born doctor with an expensive astronomy

---

* George Eastman's Kodak, a byword for cheap and cheerful snaps in pre-digital days, began life as the Eastman Film and Dry Plate Company in 1879.

habit and a remarkable wife – Anna Mary, a wealthy heiress and socialite who shared his passion and was happy to roll up her sleeves and get stuck into observing and laboratory work[3].

Using diffraction gratings made by Rutherfurd, the Drapers captured the first detailed stellar spectra on primitive "wet" plates as early as 1872, but it was their introduction to dry plates (courtesy of William Huggins's camera-mad wife Margaret, during an 1879 visit to London) that made all the difference. They set out on an ambitious programme to capture photographs of stellar spectra in far more detail than ever before, and were already establishing a reputation when, at the age of 45, Henry's life was cut short by pleurisy following a hunting expedition to the Rocky Mountains.

Thanks to Anna Mary Draper, however, the project would live on. Working with Edward C. Pickering of Harvard College Observatory, she ploughed her inheritance into the Henry Draper Memorial, a nineteenth-century spaceshot intended to catalogue the entire sky in unprecedented photographic detail. Stellar spectra would continue to play a key role, and thanks to an ingenious scheme of Professor Pickering's, they began to be captured at an ever-accelerating rate.

Pickering's bright idea was to place a carefully angled prism in front of a telescope's main objective lens, with a camera attached to the rear instead of the eyepiece. This meant that light from every star in the telescope's field of view was dispersed, so hundreds of spectra could be captured onto a photographic plate, rather than just one at a time*.

---

* A slit was unnecessary – since each star was effectively a point source of light rather than an extended object, overlap was not an issue: the natural drift of the stars across the field of view during the exposure controlled the height of the spectral bands.

The project rapidly began to accumulate thousands of stellar spectra – far more than Pickering or his male research assistants could handle. But he had a secret weapon to hand in the form of a cadre of female researchers, patronisingly referred to by some as "Pickering's Harem". Today, these remarkable women are better known by a name that swerves to the impersonal – the Harvard Computers*.

The germ of the idea for a female research team had already been sown in the late 1870s, when Pickering's wife pointed out that the family maid, a Scottish immigrant single mother called Williamina Fleming, was something of a smart cookie. Pickering took Fleming on for some light observatory admin, then set her to work organising star catalogues. Soon convinced of the advantages that women could offer†, he began to recruit more widely, often among the relatives of the observatory's male staff. By the time the Draper Memorial was set up, Fleming's fast and methodical working style made her the obvious choice to lead the group of ladies doing the hard graft of converting lines on a spectrum photo into specific wavelengths in a chart.

Just as the improvements pioneered by Draper and Pickering made it possible to capture spectra in far more detail, they also made it clear that Secchi's rather simple classification scheme wasn't really up to the job. Disagreements flared over

---

* We'll meet several of the Computers in later chapters, but for a complete account of their work, Dava Sobel's *The Glass Universe* is a recommended read.

† Cynics have noted that women gave Pickering more bang for Mrs Draper's buck, since they could be employed at half the cost of a male researcher. It's less often said that Pickering seems to have been deeply committed to bringing women into science, offering a rare route for graduates from colleges such as Radcliffe and Vassar to engage in proper research and steer their own work.

a replacement, however. Pickering and Fleming favoured an alphabetical scheme based on the amount of hydrogen that appeared to be present. However, Antonia Maury, Draper's own college-educated niece, argued that they were overlooking another important element of the spectra – the way the *width* of lines can vary from star to star – and came up with a complex classification scheme of her own that attempted to take this into account.

Maury quit the observatory in 1891*, but five years later her successor, Annie Jump Cannon, expressed similar doubts about the simple alphabetical scheme. Cannon had joined the project to classify the Southern Hemisphere stars that were now being photographed for the catalogue. Profoundly deaf from later childhood, she maintained a sunny personality that made her popular with colleagues, and was also a fast and accurate worker (classifying some 350,000 spectra during her career).

Cannon came up with a compromise that, partly by chance, foresaw many later developments in astrophysics. Rather than struggle with many different lines, she concentrated on the intensity of the Balmer series, a single set of lines linked to hydrogen. Having reordered Fleming's alphabetical letters from strongest to weakest, she then hacked away most of them to leave only the most significant differences. The surviving letters formed a fairly simple sequence: O, B, A, F, G, K and M.† Finally, she appended numbers from zero to nine

---

* She subsequently played hardball with Pickering, whose enlightened views did not always extend to crediting research assistants of either gender, to ensure her name appeared on the observatory's 1897 catalogue of bright stellar spectra.

† This is the final form – at first, Cannon also included "N" at the end, which she soon dropped.

to give a simple indication of the evolution from one letter to the next – and later astronomers have added plus and minus signs to indicate even finer divisions. So, to return to Aldebaran, its classification as a K5+ star suggests that it has very faint Balmer lines, and places it just over halfway between a K0 and an M0 star.

Putting aside its justification in the minutiae of hydrogen lines, the real reason Cannon's scheme caught on was because it cleverly reflected a far more intuitive stellar property – colour. O and B stars were without exception blue, A and F white, G broadly yellow, K orange and M red in a smooth, rainbow-like progression that would have made Newton or Fresnel proud.

And the link between spectral type and colour had even deeper implications, since by now it was widely recognised that the balance of colours in a star's continuum of light was due to an effect called "black body radiation". This strange-sounding phenomenon, described by Kirchoff as early as 1859, simply describes the radiation (both visible and invisible) given off by an object whose surface does not reflect light – something that is as true of a star as it is of the pitch-coated metal balls Kirchoff used in his laboratory.

Put simply, the rules of black body radiation mean that a star's light output is distributed across a distinctive spread of wavelengths around a central peak determined by it surface temperature: the hotter the star's surface, the steeper the curve and the shorter the wavelength of peak emission. Thus, the coolest stars emit a broad spread of radiation peaking around the red end of the spectrum (with large amounts of their energy being released in low-energy infrared waves whose wavelengths are invisible to our eyes). The hottest stars, by contrast, emit a narrower spread of radiation peaking in the blue and extending into the high-energy (and equally invisible) ultraviolet*.

Thanks to this series of links, Annie Jump Cannon's ingenious scheme (today known as the Harvard spectral classification) is a true astronomical Swiss-army knife. Spectral type, colour and surface temperature of stars become broadly interchangeable, so knowing that Aldebaran's spectral type is K5+, we could guess at its orange-red colour even if we could not see it, and predict its surface temperature of around 3,500°C.

Small wonder, then, that to this day the mnemonic "Oh, Be A Fine Girl†, Kiss Me!", a handy aide-memoire for the final order of the Harvard classifications and the colour and temperature of the stars, is one of the first things the sergeant drills into you in astronomy basic training.

---

* This also explains why there are no green stars – the temperatures where a star's light output peaks in the green part of the spectrum are also those where the light is most evenly spread out, so the combined starlight appears white.

† Or Guy, we're not fussy.

4 – MIZAR
(AND FRIENDS)

A quick waltz among multiple stars

✳

Stars, like policemen, often come in pairs. They're gregarious by nature, and the only thing your average star likes more than pairing up is hanging around in small groups, like surly teenagers kicking their heels on a celestial street corner*. Everyone remembers that bit from the original *Star Wars*, where a windswept Luke Skywalker gazes off into the twin sunsets of the planet Tatooine while the London Symphony Orchestra does its thing. Thanks to sci-fi movies, most people are quite familiar with the idea of binary and multiple stars, but like most of the stuff we now take for granted about the Universe, someone had to work it out in the first place – and a bright star in the constellation of Ursa Major was key to that story.

Mizar is another star that's easy to spot with the naked eye,

---

* Having said that, the old assumption that singleton stars like the Sun are actually in the minority no longer seems to hold true – as telescopes have become more powerful and revealed more details of the countless faint red dwarfs that outnumber all other stars in our galaxy, they've shown that these stars are overwhelmingly single. Curiously (and for reasons we don't entirely understand) it's bigger and brighter stars that tend to be multiples.

though it gets a lot more rewarding if you have binoculars or a small telescope at hand. With its close companion Alcor, it forms probably the most famous double star in the sky, but the system's true nature is far more complex than that, and each new generation of telescopes just adds to the complications.

Mizar is the fifth brightest star in the Great Bear constellation at magnitude 2.2, and a permanent resident of the sky across much of the Northern Hemisphere. Sitting neatly in the middle of the Bear's tail (the handle/pan of the famous Plough/Big Dipper), it is highest in the evening sky around June, and sinks closest to the northern horizon in December. If you live south of about 35°N, Mizar disappears completely for part of each night, but on the flipside, it also pops up over the northern horizon for Southern Hemisphere stargazers down to 35°S on winter evenings.

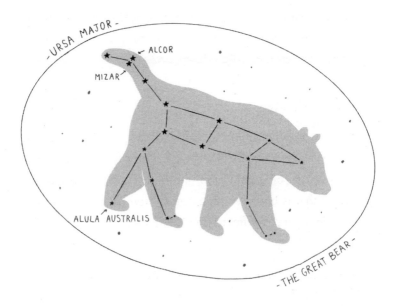

Even a casual glance at Mizar under a dark sky should show that something's up. The nearby star Alcor sits about 12 minutes of arc* (roughly one-third of a full-Moon-width) to the northwest of its brighter neighbour, at about the 10 o'clock position if you're looking at the constellation the "right way up". At magnitude 4.0, it's a fairly decent naked-eye star in its own right, but it has tended to suffer by association. In fact, Arab astronomers called it *suha*, 'the neglected one'.

The tendency among Islamic astronomers to undersell Alcor even went as far as claiming that the ability to distinguish Mizar's companion was a good test of eyesight. However, Japanese stargazers were probably closer to the mark when they named it *jumyouboshi*, the "lifespan star"– their traditions deemed that a person who *couldn't* spot Alcor was likely to die by the year's end. An exaggeration, but maybe closer to the truth than the Arabic version.

Other traditional names for Mizar and Alcor also show that the pairing was widely recognised – for instance Arab and later European astronomers knew them as the horse and rider.

Doubling down

As we've already suggested, the sky is full of double stars, and if Mizar happens to be out of view from your location, the constellation of Capricornus, the rather unlikely goat-fish hybrid that is best seen on evenings from July to October provides a couple of interesting alternatives.

---

* Quick reminder – a minute of arc is 1/60th of a degree, while a second of arc is 1/3600th of a degree.

Algiedi or Alpha Capricorni is a line-of-sight double star (that is, two stars that are only coincidentally close together as seen from Earth) with complications. Two yellow stars of magnitudes 4.3 and 3.6, designated Algiedi Prima and Algiedi Secunda respectively, are separated by about seven minutes of arc, presenting a slightly tighter and tougher challenge for the naked eye than Mizar and Alcor. In this case the two main stars really are far apart – Prima at a distance of about 870 light years and Secunda at a mere 102 light years from Earth respectively.

Each of these stars is itself a multiple system – Secunda A is orbited by a tight binary system designated B and C; with about half the separation of Mizar's two bright stars, it's a challenge for small telescopes. Prima, meanwhile, has a very tight companion that was discovered thanks to data from the Hipparcos satellite. Though relatively bright, Prima B is too close to its parent star to separate through any amateur telescope.

Beta Capricorni or Dabih is a system of at least five stars around 330 light years from Earth. Its two main components, with magnitudes 3.1 and 6.1, are easily split with binoculars. The brighter of these can be separated by the most powerful telescopes into a bright orange giant and a fainter blue star, with an unseen companion of its own. The fainter of the two main components is also a double, and there are two other possible members of the system nearby.

So, Mizar and Alcor are close – but they're not *that* close. If the 6,000-odd naked-eye stars were scattered more or less randomly across the sky, you'd expect to get a couple of close alignments like this by chance. Things get more interesting, however, when you look at the system through binoculars or a small telescope.

Mizar and Alcor through binoculars:

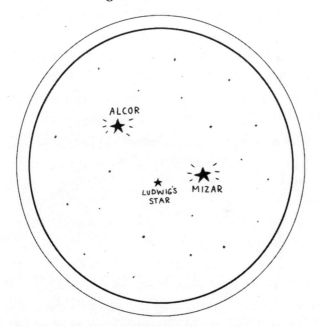

The first thing you'll spot is a third, sub-naked-eye star popping into view, roughly equidistant between Mizar and Alcor. Shining at magnitude 7.6, this star is catalogued as HD 116798, but also bears the name of *Sidus Ludoviciana*, or Ludwig's Star[*].

---

[*] Discovered by German astronomer Johann Georg Liebknecht in 1772, who briefly claimed it was a new planet and named it after his patron, the Landgrave Ludwig of Hesse–Darmstadt.

However, this is thought to be nothing more than a background star, unrelated to either Mizar or Alcor.

Get your focus nice and sharp by concentrating on Alcor, then turn your attention to Mizar. You should be able to spot that there's something wrong with it – depending on your eyesight it'll either look slightly blurry, or you may be able to see that it's actually split into two stars of differing brightness. The angle between the two is 14 seconds of arc – a small telescope should split the two stars easily.

Mizar's true nature was first spotted by Benedetto Castelli, a Benedictine monk and mathematician. Castelli was a former pupil and long-standing friend of the great Galileo Galilei, and became one of his chief allies during his famous spot of bother with the Pope. Well before all that, however, Castelli wrote to Galileo in January 1617, suggesting that his mentor should take a good look at Mizar: "It's one of the most beautiful things in the sky, and I don't believe that in our pursuit one could desire better."

Galileo's record of his own observation is probably the undated one that survives among his papers in Florence's Biblioteca Nazionale, but for some reason both Castelli's discovery and Galileo's observation were overlooked for the better part of three centuries. Another Italian, the Jesuit priest Giovanni Battista Riccioli, ended up getting the credit for spotting Mizar's double nature, thanks to a passing mention in a treatise of 1651.

Over the next century or so, Mizar became a firm favourite for stargazers, and countless other tight stellar pairings were found across the sky. But strangely, no one seems to have considered exactly what these discoveries meant. Most stargazers assumed that chance could account for the occasional close alignments they observed, but a man called William Herschel thought otherwise.

* * *

Although Herschel is best known for his 1781 discovery of the planet Uranus, this German-born astronomer had ambitions that stretched far beyond the solar system – for one thing, he wanted to make a map of the Universe.

Today, this sort of project would involve dozens of astronomers from around the world, collaborating over the Internet to use giant mountain-top telescopes. Herschel had his sister Caroline taking notes, a home-made telescope, and an ongoing obligation to spend several nights a week entertaining the trendsetters of Georgian Bath as organist at a fashionable concert hall. Fortunately, Caroline was a hugely talented and dedicated observer in her own right, and thanks to years of research and months of painstaking construction, Herschel's telescope was pretty much the best in the world.

Herschel's mapping project fuelled a growing suspicion that double stars were far more common than anyone had previously suspected – too common to be the result of random chance. In 1802, he wrote a paper for the Royal Society cataloguing 500 new nebulae and star clusters, and raised the prospect that tight double stars might be bound together by gravity[1]. He proposed that such physical pairs (as opposed to line-of-sight or "optical" doubles) might be referred to as binary stars.

Herschel's argument from statistics was convincing, but some astronomers still had their doubts. A couple of years later, however, he found undeniable proof that binary stars were real. In May 1780, he had spotted that another of Ursa Major's stars, known as Alula Australis (in the bear's hindmost foot), was also double – a beautiful, quite closely matched pair of yellow-white stars of magnitudes 4.3 and 4.8 in a roughly north-south alignment. Yet, when he looked again at the star in 1804, he found the stars' alignment was closer to east-west. They had shifted their position significantly over 24 years, and Herschel was able to announce the first evidence for two stars in orbit around each other[2].

Once Herschel had shown that Alula Australis was a bona fide binary star system, the search was on for others, and Mizar was a prime target for investigation. Sadly, it soon became clear that in this case the two stars Mizar A and B followed a much longer orbit, and even observations decades apart showed no signs of change (according to current estimates, the pair take about 5,000 years to complete a single sluggish circuit).

Fortunately, there was another way of confirming that the two stars of Mizar were generally associated in space – the fact that they showed an identical parallax shift (see 61 Cygni if you're in need of a reminder). The confirmation had to wait for almost half a century, but it was finally accomplished by German astronomer Ernst Klinkerfues at Göttingen Observatory in the early 1850s. Klinkerfues showed that the stars shifted their position back and forth by a tiny 0.043" over each year (the width of a full stop on this page at a distance of about 500 metres). This suggested that both stars lie about 76 light years from Earth (not far off the modern measurement of 82.2 light years).

Parallax measurements of Alcor seemed to indicate that it was a few light years closer to Earth, leaving the question of its physical relationship with Mizar unresolved. However, in 1869, British amateur astronomer and popular astronomy writer Richard Anthony Proctor reported a remarkable discovery: six of the bright stars in the Plough or Big Dipper, including both

Mizar and Alcor, are moving through space at more or less the same speed and in the same direction*.

Today, astronomers call Proctor's discovery the "Ursa Major Moving Group"[3]. Fourteen core members have been identified across the Great Bear and bordering constellations; along with several dozen other stars, they are now all thought to have originated in the same cluster of newborn stars around 500 million years ago.

* * *

As if Mizar hadn't already secured its place in the panoply of great stars, 1889 saw it play a pivotal role in opening a whole new era of astronomy, thanks to a remarkable discovery at Harvard College Observatory.

We saw in the previous chapter how William Huggins had claimed to calculate the motion of Sirius by analysing a shift in its spectral lines, but the photographic methods pioneered by Edward Pickering at Harvard turned out to be far more suited to this kind of job.

While detecting the "proper" motion of stars (sideways movement against the background of the celestial sphere) takes little more than patience, "radial" motion (movement towards or away from Earth) demands a different approach. Christian Doppler had realised as early as 1842 that light from approaching or receding objects would shift towards the blue or the red end of the spectrum as its waves were compressed or stretched. Doppler had hoped this might explain the varied

---

* For the record, the exceptions are Dubhe, the northernmost of the two "pointer" stars on the right-hand side of the bowl, and Alkaid, the star on the end of the dipper's handle.

colours of stars, but hadn't really appreciated the vast speeds required to affect a star's colour. In cases where stars moved at a few kilometres an hour, the Doppler shifts were far smaller and best detected by measuring shifts in the precise wavelengths of absorption lines in a stellar spectrum.

Pickering's team photographed spectra from Mizar and nearby stars on 70 separate nights between 1887 and 1889. Antonia Maury was tasked with analysing them, and soon noticed something strange about Mizar A: At times, the prominent dark "K" line, created by calcium in the star's atmosphere absorbing energy, looked fuzzier than it did in other stars. On three of the spectra it had split completely into two separate lines with wavelengths to either side of where the K line should normally be. This "doubling" seemed to come every 52 days, and further spectra soon confirmed the cycle.

What kind of situation could cause a star's light to split apart and reunite in a regular cycle? Pickering realised that, despite its appearance as a single star through even the most powerful telescopes, Mizar A must itself be a binary pair comprising two similar stars in a tight orbit. The speed and direction of each star's motion relative to Earth is constantly changing, so that at any moment their combined Doppler shifts are tugging one set of spectral lines towards the blue end of the spectrum and the other towards the red. When the effect is at its greatest (with one star moving more or less directly towards Earth and the other directly away), these blue and red shifts are strong enough to pull the two sets of spectral lines apart completely, producing the doubling effect. At other times, when the stars move more or less "sideways" relative to Earth, the Doppler shift is unnoticeable.

Pickering recognized at once that a system like this could

tell you a tremendous amount about the properties of the stars involved. Based on the fact that the Doppler effect seemed to be affecting light from both stars equally, he imagined a simple model of the Mizar A system – two stars of equal mass, following circular orbits around the system's shared centre of mass or "barycentre" like kids on opposite sides of a merry-go-round, with 104-day periods on a plane that neatly lined up with Earth. Based on the strength of the Doppler shifts, he calculated that the stars were moving along their orbits at roughly 160 kilometres per second, with a separation of about 230 million kilometres – about the average distance between the Sun and Mars.

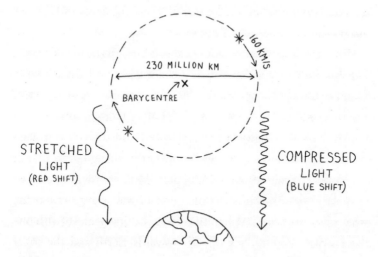

So far, so straightforward – but Pickering realised you could take the calculation a step further. The speed of any object along its orbit depends not just on the size of that orbit, but also on the mass of the object that it's orbiting. In our solar system, for example, Mars takes 687 days to orbit the Sun, but if the Sun was heavier, Mars would have to orbit faster. Based on this principle, Pickering estimated that the Mizar A system's mass must be equivalent to at least 40 suns[*], evenly shared out at an impressive 20 solar masses per star.[4]

Pickering's discovery was widely lauded as a feat of astronomical derring-do and a new way of probing the properties of stars. Mizar became the prototype for a whole new class of stars known as spectroscopic binaries, yet it soon began to misbehave. Sometimes, the lines were double as expected, sometimes they appeared fuzzy, and more often than not, they remained sharply defined and distinctly singular even when the model predicted they should be broadening. Pickering tried various "fixes" for his model – halving the orbital period, introducing stretched elliptical orbits, and even adding a third unseen body to the system.

But it eventually fell to a rival spectrographer, Hermann Carl Vogel of Germany's Potsdam Observatory, to sort out the mess. Observations in the spring of 1901 managed to capture Mizar's spectrum almost continuously across five weeks, revealing that the star's cycle actually repeats in a much faster 20.5 days, with the maximum blue and red shifts just four days apart.

Vogel's measurements showed him that Mizar A's stars are both following highly *elliptical* orbits, accelerating when they were close to the system's barycentre at one end (to produce the greatest red and blue shifts) and slowing around maximum separation[5]. The gap between the stars varies, according to modern measurements, between 16 million and 54 million kilometres

---

[*] And this was a *minimum* – if the orbit was inclined to Earth (so that the motion revealed in the Doppler shifts was only a fraction of the overall velocities), then the stars could be much heavier.

(although a sketch diagram makes it look like the orbits overlap, their tilt in different planes, which is a little hard to show in an overhead sketch, ensures they don't collide in a Scalextric smash-up of cosmic proportions).

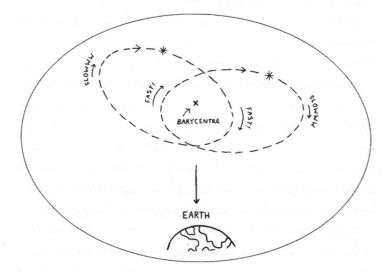

Vogel's solution reduced the masses of the two stars at a stroke. Instead of maintaining 160 kilometres per second *throughout* a very long circular orbit, the stars only had to reach this kind of speed around close approach. Their *average* speed could be much lower, and this in turn suggested that the two stars actually had much less daunting masses – each around 2.5 times that of the Sun*.

The discovery of spectroscopic binaries immediately revealed that there were many more stars in the sky than anyone had previously suspected. In order to be identified through a telescope,

---

* These days, Pickering's super-heavy stars would quickly set alarm bells ringing. We now know that stars with tens of times the Sun's mass are rare and very distinctive (see Eta Carinae, for example) – and Mizar really doesn't fit the bill.

a so-called "visual binary" has to have a significant angle between its stars – meaning it must either be close to Earth, or separated by a truly huge distance. Even with modern telescope technology, this puts a limit on the number of visual binaries we can detect. Spectroscopic binaries, on the other hand, can be identified in light from the most distant stars, and they were soon springing up everywhere.

Not to be outdone, in 1890 Vogel discovered a second type of spectroscopic behaviour while studying a famous variable star called Algol (not to be confused with Alcor). Although he couldn't detect the line-broadening and doubling behaviour seen in Mizar A, he could see that the star's spectral lines wandered back and forth in a period that matched its changing brightness. This was clinching evidence that Algol, too, was a binary – one in which most of the light is produced by just one of its component stars.

Vogel's "single-lined" spectroscopic binaries turned out to be far more common than Pickering's double-lined versions, though sadly they're rather more limited in what they can tell us about star systems. In 1908, America's Edwin B. Frost and Germany's Hans Ludendorff independently discovered that Mizar B was a binary of this type, and so Mizar went from being the first triple star system to be discovered, to the first *quadruple* star. Today, it is generally agreed that this fainter pair have masses around 1.6 times that of the Sun.

* * *

Despite all the attention paid to Mizar, its fainter sibling Alcor has remained frustratingly hard to pin down. Though clearly a member of the Ursa Major Moving Group, its relationship to

Mizar has remained uncertain, and the star itself has been equally badly behaved.

Alongside his 1908 discovery that Mizar B was a spectroscopic binary, Frost published evidence for unpredictable changes in Alcor's light, which he thought might be caused by an unseen binary companion tugging on the visible star. Canadian astronomer John F. Heard reported similar measurements in 1949, and a couple of generations of astronomy books persisted in describing Alcor as a spectroscopic binary in its own right, although the idea was actually disproved as early as 1965.

Yet, just as astronomers had readjusted to the idea of Alcor as a lone star, there was another twist in the tale. In 2009, Alcor was included in a planet-hunting survey using a highly sensitive infrared camera called Clio. Ignoring most of the higher-energy light pumped out by stars like Alcor, this camera swept up the longer wavelengths of radiation emitted by much cooler objects. And there, just one second of arc away from Alcor in every one of the 129 exposures, was a faint red dwarf companion star[6].

This new discovery may help solve a long-standing puzzle, namely the source of high-energy X-rays detected by an orbiting satellite from the direction of Alcor in the 1990s. Stars like Alcor don't usually create these high-intensity outbursts, but perhaps surprisingly, much fainter red dwarfs *do* (just one of the ways in which these tiny stars can pack a surprising punch – see Proxima Centauri). Perhaps Alcor's X-rays are actually coming from its companion star?

The existence of Alcor B may also have finally settled the old argument about Alcor's true relationship to Mizar. While recent parallax measurements place Mizar and Alcor as little as one-third of a light year apart, doubts remained because Alcor's motion seemed to differ from the path expected for a star orbiting Mizar. The influence of Alcor B, causing Alcor A to wobble slightly on its long orbit, neatly explains this difference, and shows that the Mizar and Alcor systems really are almost certainly bound together by gravity, orbiting each other in a million-year slow dance.

★ ★ ★

In Mizar's long history, it's ticked off a series of remarkable firsts – it was the first widely recognized stellar pairing or optical double, the first close visual binary recognised in the age of the telescope, the first to be photographed, and the first spectroscopic binary. Along the way it's helped astronomers to understand the properties of stellar orbits and to weigh the masses of stars. Today, we can acknowledge it as one of the most complex star systems in the sky – and given its track record, would anyone want to bet that it's given up all its secrets just yet?

5 – ALCYONE AND HER SISTERS

How the sky's most beautiful star cluster
inspired a very important diagram

※

*I*f the stars in the sky can vary significantly in both colour and brightness, then are each star's precise mix of properties just a throw of the dice, or are some combinations more common than others? Discovering the answer to this question in the early twentieth century would point the way towards understanding the life cycle of stars, and reveal some neat tricks that allow astronomers to estimate the distance to remote parts of the galaxy. The story starts (and finishes) with the Pleiades – the most famous star cluster in the sky.

The Pleiades mark the shoulders of the great charging bull constellation Taurus, and are an unmistakeable sight in the night sky throughout the months of northern winter. A distinctive fishhook-shaped blob of light, they lie to the north and west of Taurus's head (itself marked by bright orange Aldebaran and the v-shaped, slightly looser Hyades cluster). In the Northern Hemisphere, look for them to rise over the eastern horizon in the hours before sunrise from July onwards, moving into the evening sky from October as heralds of the coming winter, and finally disappearing into the sunset as they draw close to the Sun in April. South of the Equator, they skim across the northern horizon from northeast to northwest in the same months.

Thanks to their prominence, the Pleiades are at the centre of many myths and legends from around the world. The first-millennium civic planners responsible for Teotihuacan in present-day Mexico designed their entire city so its avenues aligned with the location where the Pleiades set over the mountains. They are shown alongside the Sun and Moon on the mysterious Nebra Sky Disc (dug up by treasure hunters in northern Germany in 1999 and dated to around 1600 BCE), and they are probably depicted on a Lascaux cave painting of a charging bull from 18,000 years ago.

Western stargazers have inherited the ancient Greek tradition of the stars as the Seven Sisters, a group of demigoddesses charged with raising Dionysus, god of fertility and wine. Later, they were chased by the hunter Orion for reasons we won't go into, until Zeus (in a somewhat hypocritical stand against sexual harassment) took pity on them in the way that only a Greek god can, transforming them into doves that flew into the sky to escape the pursuing mythological sex pest. As Greek myths go, theirs is definitely in the second rank, but that didn't prevent them prancing in diaphanous gowns through the overheated imaginations of many Victorian painters.

Despite the Seven Sisters' nickname, it's often said that only six stars are visible to the naked eye for people with average eyesight. A few sharp-eyed stargazers have reported eight, 11, and even as many as 16 stars with the naked eye – the main challenge is not so much their brightness, as the fact that the closely packed stars tend to blur together.

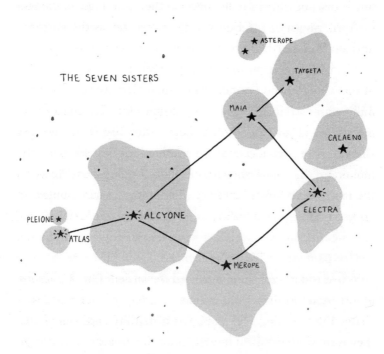

THE SEVEN SISTERS

TAURUS

The nine brightest stars are named, in no particular order, after all seven sisters *and* their parents Pleione* and Atlas (the strongman doomed to hold up the heavens). Alcyone is the brightest, while in descending order the others are Atlas, Electra, Maia, Merope, Taygeta, Pleione, Celaeno and Asterope. Look at the cluster on a good night (or through a pair of binoculars) and

---

* Pleione is probably a name invented "after the fact" to explain the pre-existing name Pleiades. Some etymologists think the cluster's name actually originates from the Greek word for sailing – the idea being that their first appearance before sunrise in the morning sky signalled the onset of calmer seas and the beginning of the Mediterranean boating season.

you should be able to make out its shape as a rough rectangle with outlying stars – Alcyone is the one in the northeast corner of the rectangle.

★ ★ ★

The Seven Sisters were an obvious target for the first telescopes, and as early as 1610 they made it into Galileo's bestseller *The Starry Messenger* alongside such crucial discoveries as the satellites orbiting Jupiter and the craters of the Moon. Galileo mapped out dozens of previously unknown stars, bringing the group's total membership up to 36. Most people who thought about it realised that such an alignment occurring by chance was very unlikely, but it was not until 1767 that English clergyman and philosopher John Michell crunched the numbers and announced that the odds of a chance alignment were one in half a million.

By Michell's time, telescopic improvements had led to the discovery of many other star clusters and fuzzy patches of light in the night sky. These clusters and nebulae (from the Latin for clouds) were intriguing objects in their own right, but for French stargazer Charles Messier, they were a downright nuisance. Messier was more concerned with scanning the sky to discover comets – visitors from the outer solar system that frequently appeared as… fuzzy patches of light in the night sky.

In 1771, Messier published a handy catalogue of comet imposters to save himself and others further confusion, with the Pleiades, perhaps unnecessarily, included as object M45. In one of life's little ironies, the Messier Catalogue became his lasting legacy, and has been a handy shortlist of some of the sky's most beautiful objects for generations of astronomers, both amateur and professional.

The Pleiades have always held a near-magnetic attraction for astronomers, and have probably been studied more than any other stars in the sky. The idea that they were a physical cluster was intriguing, and many stargazers tried to work out whether they were orbiting each other or moving through space. In 1846, German astronomer Johann Heinrich von Mädler, director of the Dorpat Observatory (in modern Tartu, Estonia) concluded from measuring the proper motions of 3,000 stars that our solar system and everything else revolved around a central concentration of mass in the Pleiades. Alcyone therefore enjoyed a brief and glorious moment as the central sun of the entire Universe (other galaxies hadn't yet been discovered). Mädler's measurements were soon surpassed, but in some ways he had the right idea, as we'll see when we get to the star known only as S2*.

The Sisters were hugely useful to astronomers trying to get to grips with stellar properties in the early twentieth century. The cluster was too far away to measure its distance directly using parallax (see 61 Cygni), but on the other hand, because it was undeniably a physical group, all of its stars could be considered as lying at effectively the same immense distance from Earth. This meant that, for once, you could stop worrying about how the varying distances to stars affected their apparent brightness, and assume instead that differences in the apparent magnitude of cluster stars in the night sky were a direct reflection of differences in their true brightness, diminished by the same distance factor in every case. An ingenious young Dane called Ejnar Hertzsprung soon found a way of putting this to good use.

---

* Even today, Alcyone's moment in the limelight has not entirely been forgotten by the more woo-woo parts of the Internet, as a quick search for the term "central sun" will show.

By the early 1900s, the Henry Draper Memorial project (covered in some detail under Aldebaran) was in full swing, with thousands of stars now being catalogued according to Annie Jump Cannon's ingenious "spectral type" classification – an indicator of both surface temperature and colour.

It was clear that colour and luminosity (a star's intrinsic brightness) were independent of each other, but might the absorption spectrum also hold a clue to a star's luminosity? The possibility was enticing, and one of the keenest minds pursuing this idea was Hertzsprung – a trained chemist who had lately returned to his boyhood passion for the stars. Largely self-taught, Hertzsprung had a flare for asking the right question at the right time – and among many other things he was fascinated by the variations in the width of spectral absorption lines that had so troubled Antonia Maury.

Hertzsprung came up with an ingenious rule of thumb that allowed him to put stars into broad groups based on their likely distance from Earth. Remember how earlier astronomers identified potentially nearby stars by looking for those with the biggest proper motions across the sky? Hertzsprung turned the idea on its head, noting that if all stars are moving through space with roughly similar average speeds, then their perceived proper motion across Earth's sky will get smaller the further away they are.

Using proper motion as a crude indicator of distance allowed Hertzsprung to guesstimate distances for many of the stars that Maury had previously catalogued. He soon found that stars with narrow absorption lines tended to show smaller proper motion than stars with the same spectral classification but broader lines. The narrow-lined ones, it seemed, were more distant than the broad-lined ones, and since this apparently had no systematic effect on their average brightness in Earth's skies, they must be intrinsically more luminous.

By around 1905, Hertzsprung felt confident that he had identified a fundamental division between two classes of stars – narrow-lined, highly luminous ones and broad-lined, less luminous ones. Within a few years, astronomers began to call them giants and dwarfs – often attributing the terminology wrongly to Hertzsprung himself, who actually felt it was misleading.

Searching for a way of comparing properties between large numbers of the less-luminous stars, Hertzsprung soon focused on the two great star clusters of Taurus – the Pleiades and Hyades. By around 1908, he had devised a graph to compare the magnitude of stars in the Pleiades with their spectral type. However, other projects, the demands of his chemical work and the need to gather more data meant he was not ready to publish his comparison until 1911.

Hertzsprung's final diagram[1] plotted the apparent magnitude of stars along the horizontal axis, with the brightest (such as Alcyone) on the left and increasing numbers of fainter stars towards the right. The vertical axis, meanwhile, indicated colour and surface temperature, with hotter, bluer colours at the bottom.

The graph was limited in its scope, largely because the stars of the Pleiades skew heavily to the blue end of the spectrum (being nearly all B and A stars in Annie Jump Cannon's scheme). Nevertheless, there was still enough variety to show a clear pattern – the most luminous stars were also the hottest, and as brightness dropped so did temperature.

Hertzsprung's diagram was the first version of what we now

call a Hertzsprung–Russell diagram (H–R for short)[*2]. It was clearly hinting at something very important about the stars, and the young newcomer's mentor, respected German astronomer Karl Schwarzschild[†] was convinced it was a major discovery. Hertzsprung wrote several times to William Pickering at Harvard, arguing for the reinstatement of Maury's line-width classes in future catalogues, but Pickering remained unconvinced. It would take someone else to make the breakthrough and transport the Dane's ideas into the mainstream.

That someone turned out to be Henry Norris Russell. The son of a Presbyterian minister from New York State, he had inherited a talent for maths from his mother Eliza, and unlike Hertzsprung, he had pursued his interest in the stars along a more straightforward academic route. A degree and PhD at Princeton were followed by research at Cambridge University around 1902–5, where he focused on ways of determining the properties of binary stars.

Binaries offered another way of approaching the same problem as clusters – finding and comparing the properties of stars at an identical distance from Earth. In addition, however, Russell realised that you could use a binary's orbit as a lever to prise open its true distance. The principle was relatively simple, and an extension of all the arguments about Mizar we saw in the last chapter: if you could find out enough about a binary system to calculate the true dimensions of its orbit, then you

---

[*] Ejnar gets the credit largely because he seems to have had the idea first, but he was pipped to the post by a largely forgotten German astronomer called Hans Rosenberg who published an almost identical chart using his own system of spectral properties in 1910.

[†] More on him when we come to Cygnus X-1.

could compare that with the *apparent* size of its orbit in the sky and figure out how far away it was. The method only worked for a few stars, but it got Russell interested in the whole question of comparing stellar properties such as size, mass, spectral type and luminosity.

Freshly minted as a Princeton professor in 1911, he was therefore well placed to see the potential of Hertzsprung's idea, but realised the need to extend the range of spectral types beyond a single star cluster could offer. But here there was an obvious hitch. While there were plenty of other clusters that lent themselves individually to the same trick that Hertzsprung had used, and a few score stars whose luminosities could be worked out directly through the parallax method, how could you combine them onto a single diagram?

As sometimes happens, Russell simply got lucky, as two further methods of estimating the true distance of stars had just been identified: the "secular parallax", devised by Dutchman Jacobus Kapteyn and the "moving cluster" method created by American Lewis Boss. Both methods worked only for star clusters, but were ideal for Russell's purposes. And happily, they had both already been applied to the Hyades, whose large size and relatively loose scattering of stars led many to suspect it was the closest star cluster to Earth*. The results seemed to bear this out, delivering near-identical figures that averaged out at a distance of 135 light years.[3]

Russell unveiled his extended diagram at the American Association for the Advancement of Science meeting in Atlanta, Georgia at the end of December 1913, and wrote it up in a

---

* Overlooking the Ursa Major Moving Group discussed in the previous chapter.

detailed review the following year[4]. Altogether, he was able to scrape together parallax measurements for around 300 stars of all spectral types, but the lack of a directly measured distance forced him to leave out the Pleiades.

This new version of the H–R diagram set a template that has been followed ever since, with a measure of luminosity[*] on the vertical scale, and spectral type on the horizontal. The pattern hinted at in Hertzsprung's diagrams was suddenly crystal clear – a large majority of stars (which Russell happily described as dwarfs) lay along a diagonal line linking luminous blue stars to faint red ones – a strip that Hertzsprung had already named the "main sequence". Meanwhile, the giants ran horizontally across the top of the diagram in a broad band. The separation between the two groups of stars was greatest at the red end of the spectrum, where

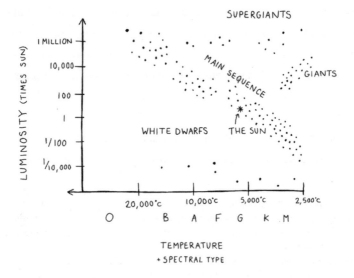

---

* In this case, "absolute magnitude", a scheme invented by Kapteyn that calculates a star's brightness if we were seeing it from a distance of 10 parsecs of 32.6 light years.

stars were either extremely luminous or very faint, and nothing of similar brightness to the Sun could be found.

Russell's diagram might have been a huge leap forward, but he was aware of its power to mislead. In particular, he knew the relative numbers of stars must be thrown out of balance by an unavoidable bias in the stars that we can see. Giants shine out over many hundreds of light years, while dwarfs – especially faint orange and red ones – are invisible beyond our cosmic back yard. Select a truly random sample of stars in our galaxy, and you'd find many more dwarfs and far fewer giants than the diagram might suggest. The main sequence was clearly where the vast majority of stars spent the vast majority of their lives, and this would be a key observation that future theories of stellar evolution would need to address.

An example of Russell's smart thinking with more immediate applications was the idea of turning parallax on its head. It should be possible, he pointed out, to measure a star's spectral type, identify which branch of the diagram it sits on*, and so work out its likely absolute magnitude and distance. He even provided a deceptively simple-looking mathematical formula for doing the hard work.

Unfortunately, getting *useful* distance estimates was trickier than it seemed. With large margins for error in most of the figures

---

* The mystery of the varying line widths was also resolved in 1913, when German physicist Johannes Stark discovered an effect known as pressure broadening that makes the spectral lines associated with a gas grow broader when it is under higher pressure. Because the surface of a dwarf star is more compact than that of a bloated giant, it produces broader lines. The concept was finally folded into stellar classification at Yerkes Observatory, Wisconsin in the 1940s, with the introduction of "luminosity classes" (0, I, II, III, IV or V working down from the most brilliant supergiants to normal dwarfs) alongside the Harvard spectral type.

used to plot them, both the giant and main-sequence bands were broad enough to accommodate huge variations in brightness, rendering distance estimates meaningless. Even Russell admitted that his formula only provided an *average* parallax.

It didn't take long, though, before someone figured out a way of breaking the impasse – and once again Alcyone and her many sisters would show the way forward. In 1918, William Henry Pickering (younger brother of Edward, whom we've already encountered running the Harvard College Observatory) announced he had made a plausible distance estimate for the Pleiades.

Pickering's technique[5] was a series of fairly straightforward statistical tricks. First, he subdivided the stars into groups of different spectral types, and worked out the average apparent magnitude for the stars in each type. From this, he calculated the corresponding average parallax using a variation on Russell's own formula. Finally, he simply took an average of those distances. He wound up with a distance of 656 light years – about 50% higher than the modern-day value of 444 light years [*].

One could say it's the thought that counts, but in fact Pickering had hit upon something important: the idea that the large number of stars in a cluster and the ability to average their properties makes calculations easier and more robust than those for individual stars. Pickering, like Hertzsprung, was working with data for just a few dozen stars. More recent surveys have found around 1600, stretching all the way down the main sequence through yellow Sun-like stars to faint red dwarfs, all of which add to the wealth of data and the precision with which these sorts of methods can be applied.

---

[*] Pickering didn't help himself by choosing to omit the cluster's brightest stars, believing (wrongly) that they were on the giant branch of the H–R diagram.

Within a few years, Pickering's basic concept gave rise to a method called "main-sequence fitting". The idea here is to put together an H–R chart for an individual cluster of stars using their apparent magnitudes, and then adjust its vertical position so it neatly overlies the more general H–R diagram. This will tell you the offset between apparent and absolute magnitudes in the cluster, and from there you can work out its distance.

Hertzsprung himself identified a useful refinement to this idea in 1929, when he considered the differences between the Pleiades, Hyades and another famous cluster called Praesepe (in the constellation Cancer)[6]. The stars of the Pleiades cling to the ascending line of the main sequence all the way to the top and hence the brightest members are all blue and white, but in contrast he noted that the other clusters both contain highly luminous yellow, orange or red stars, while the hottest and most luminous blue stars tail off somewhat further down the main sequence.

This discovery, now known as the "main sequence turnoff point" of a cluster, not only allowed fitting to be done more precisely and the distance to clusters to be refined, but it would also play a key role in unravelling the story of how stars evolve and die. We'll be revisiting this in later chapters, but our modern understanding is that all of the stars in a cluster take up a position somewhere along the main sequence band of the diagram shortly after they are born. They'll spend most of their lives here, obeying a simple relationship between temperature and luminosity, but the hottest and brightest ones age more quickly than their fainter siblings. As they near the end of their lives, they transform into other stars that may shine even more brightly, but have cooler and redder surfaces. In this way, the bright blue end of the main-sequence diagonal is gradually eroded away, while the numbers of bright red and orange giants increase in its place – and the older the cluster, the further the main sequence is worn down. The Pleiades, young on a cosmic timescale at an estimated 50 million years old, are so far unaffected by this stellar ageing process – but when it comes, beautiful blue-white Alcyone will be the first to feel it.

6 – The Sun

*The star on our doorstep and
what it can tell us*

Y ou could be forgiven for wondering why we've left the
Sun until this point in proceedings. It is, after all, the
most obvious star in the sky, without which none of us would be
here. From the point of view of understanding the wider Universe,
however, it's important to first put our local star in context. Now
we've done that, we can concentrate on how the Sun provides
a handy laboratory for testing our ideas about other stars, and
finding out what really makes them tick.

For instance, to understand how the Sun fits in among other
stars, we need to use the spectral classification system we saw
applied to Aldebaran. Looking at the total amount of energy it
pumps out and the way this is spread across a variety of different
wavelengths of light, we find that the Sun is a "G2 V" star. The
"G2" puts it at the hotter end of the yellow G-class stars*, while
the "V" puts it in the dwarf luminosity class. It might be hard to
imagine a 1.4-million-kilometre ball of exploding gas as a dwarf

---

* The Sun's yellowish appearance is enhanced by Earth's atmosphere, which scatters
shorter, bluer wavelengths of light on their way through the atmosphere, creating the
blue glow of the sky and leaving the Sun slightly looking slightly redder than it really is.

of any kind, but in this sense the term simply indicates that it's in the long, mature middle phase of its evolution (along with the vast majority of stars we can see in the sky).

The first and most obvious *difference* between the Sun and those other stars is that we can see its disc – every other star in the sky is nothing more than a point of light through even the most powerful telescope*. What appears to be a solid surface, however, is an illusion – the Sun's visible boundary (known as the photosphere) is actually more like a fog bank – a relatively narrow layer where the hot gas thins enough to become transparent, so that rays of light can dash off into space without further interruption. The temperature of any star's photosphere depends on both its surface area and the amount of energy trying to escape through it. In the Sun's case, this works out at about 252 megawatts per square metre, heating the photosphere to a toasty 5,500°C.

But even to the earliest skywatchers, it must have been obvious that the Sun extended well beyond the blazing photosphere – when the Moon passed in front of the Sun during eclipses and blocked the disc from view, they would have seen colourful flames rising above its surface, and tendrils of faintly glowing gas extending far beyond them before fading into the blackness of space. These days we call the ruddy flames prominences – huge arcs of superheated gas that run along magnetic rivers through a region called the chromosphere. Above these arcs lies the Sun's outer atmosphere (or corona), where gas is far more tenuous and glows with a milky-white light as it blows outwards into a solar wind that streams across the solar system.

Nevertheless, for thousands of years, philosophers, stargazers

---

* Although see Betelgeuse for a clever way around this problem.

and religious scholars considered the Sun to be a perfect light in the sky – a symbol of divine perfection from a realm immune to the change and decay of Earthly existence. And then, in 1608, Dutch spectacle-maker Hans Lippershey invented the telescope.

The first idiot to try looking at the Sun through this new invention was Englishman Thomas Harriot, an inveterate scientific tinkerer who had accompanied one of Walter Raleigh's transatlantic expeditions to Roanoke in the 1580s, and later hunkered down teaching natural philosophy. In 1609 he wrote letters recording observations through his new spyglass (referred to rather marvellously as a "Dutch trunke"). Fortunately, he had enough sense to wait until the Sun was low on the horizon and filtered by the intervening atmosphere.

Galileo tried the same thing in 1612, but soon adopted a much safer way of looking at the Sun that had been invented by his former student Benedetto Castelli. This involved projecting the Sun's image through a telescope onto a screen at a safe distance behind the eyepiece.*

Galileo thus became the first person to publish an observation of the black blobs known as sunspots, and so he's often given the credit for discovering them. In reality, we now know he was beaten not only by Harriot, but also by several medieval Chinese, Korean and European stargazers who worked without optical assistance.

Sunspots, or starspots as they're known on stars other than the Sun, appear as dark areas of the photosphere and seem to be

---

* The idea that Galileo's early solar observations were responsible for his blindness in old age is one of those myths that refuses to die – solar retinopathy doesn't usually wait for 25 years to make its effects felt. But in case it needs saying: NEVER stare directly at the Sun and CERTAINLY don't point a telescope or binoculars near it if your eye is anywhere near the other end.

a common feature of all stars, though they vary hugely in size and intensity. Although individual sunspots aren't permanent features, some can persist for several weeks, and it was through tracking their movements that Galileo first showed that the Sun must be rotating and that the spots were physical flaws in its perfection, rather than small bodies passing in front of it. The suggestion that the heavens could be flawed and imperfect caused Galileo almost as much trouble as his support for a Sun-centred solar system.

The first person to keep the Sun's appearance under regular surveillance was Danish astronomer Christian Horrebrow. Working from 1761 from a picturesque tower in central Copenhagen*, he charted the Sun's appearance, and by 1775 he had spotted hints of a cycle in both the size and number of sunspots. However, Horrebrow's research was cut short by his death the following year, and it was several more decades before this solar cycle was rediscovered by German apothecary-turned-astronomer Heinrich Schwabe in 1843. Rudolf Wolf, director of the observatory in the Swiss capital of Bern, was intrigued by Schwabe's report and took up studying sunspots, and scouring historical records, in earnest. He pinned down the sunspot cycle to an 11.1-year repeating pattern, extending back to at least the 1740s[1].

In fact, the cycle seems to be a more-or-less permanent fixture, but a prolonged century-long drop in sunspot numbers, beginning in the early 1600s, made it difficult for Wolf to trace it further back. The idea that the gap in the records reflected a real

---

* The Rundetaarm – purpose-built by King Christian IV with a 35-metre-high equestrian staircase – a wide spiraling ramp ideal for those who fancied a spot of horseback stargazing.

drop in numbers was first suggested by Germany's Gustav Spörer in 1887, but it is today known as the Maunder Minimum after Edward Maunder of London's Royal Observatory, who drew attention to it.

Spörer, who worked at both the University of Berlin and Potsdam Observatory, hasn't had much luck with posterity – two other discoveries he made in the 1860s are also usually attributed to a more famous Brit, Richard Carrington, who worked around the same time. Spörer and Carrington both used sunspots to show that the Sun rotates faster near the equator and slower at higher latitudes, confirming for the first time that it was not a solid body[*]. They also identified a gradual drift from high to low latitudes through each solar cycle: sunspots start off in small patches at mid-latitudes (around +/- 40°), develop into major outbreaks as they drift closer to the tropics, and finally peter out around the equator, before the entire cycle restarts at higher latitudes.

These discoveries proved crucial to solving the mystery of sunspots – but a final clue was still needed. In 1908, American astronomer George Ellery Hale discovered that the spectral lines in light from the spots were split and distorted in ways that suggested a strong magnetic field was present. By April 1919, Hale and his colleagues had discovered patterns that were enough to suggest a magnetic explanation for the entire cycle, so here it comes...

The Sun's interior is incapable of sustaining a permanent

---

[*] The 1769 discovery by Scots astronomer Alexander Wilson that sunspots are depressed in relation to their brilliant surroundings had led many to conclude that the Sun's light came from a layer of brilliant clouds overlying a dark, possibly solid, and perhaps even inhabited, surface.

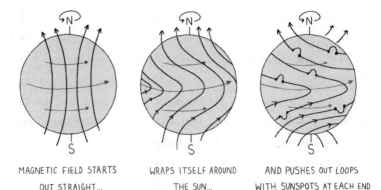

MAGNETIC FIELD STARTS
OUT STRAIGHT...

WRAPS ITSELF AROUND
THE SUN...

AND PUSHES OUT LOOPS
WITH SUNSPOTS AT EACH END.

magnetic field, but it does contain layers of swirling electrified gas that can create a temporary one. At the beginning of each approximately 11-year sunspot cycle, this field runs smoothly between one pole and the other beneath the Sun's surface, emerging near the poles a bit like Earth's own magnetic field. However, because the Sun spins faster at the equator than at the poles, equatorial regions gradually drift eastward of those at higher latitudes, stretching the magnetic field they carry within them a bit like pulled candyfloss. After many rotations, the increasingly tangled field starts to become distorted, and magnetic loops burst out through the surface like solar hernias. The points where these loops burst out and re-enter become the sites of sunspots. By lowering the density of the gas around them, they allow it to cool down so that it appears dark compared to the surrounding photosphere*.

Spots arise in matched pairs (one magnetically "north", the other "south") at either end of the magnetic loops. Over time, their numbers increase as the field winds up, but as spots from

---

* Sunspot temperatures are still around 3–4,000 °C – you can only call them "cool" in the context of the Sun's average surface temperature of around 5,500°C.

opposite hemispheres come closer to each other around the Equator, they cancel out and numbers start to fall. After about 11 years, the magnetic field fades away to nothing, only for the swirling electrified gas to do its thing again and generate a new field, this time with the magnetic poles reversed. Thus the underlying solar magnetic cycle actually takes 22 years (with some substantial give-or-take) to complete.

Sunspots might be the easiest way of tracing the Sun's changing activity, but they're far from the only one, as people around the world discovered to their alarm in 1859. On 1 September, Richard Carrington and other observers studying a large and complex cluster of sunspots were dazzled by an immense burst of incandescent light on the solar surface. Within hours of this eruption, Earth's northern and southern lights flared into spectacular life, disturbing sleep patterns around the world with a blood-red glow and providing enough light for New Yorkers to read their newspapers at midnight. Compasses went haywire, and the electric telegraph system descended into chaos as machines began to produce painful sparks*. A vast unknowable *something* had emerged from the Sun and swept past Earth, dumping vast clouds of subatomic particles into the upper atmosphere, disrupting the magnetic field and leaving confusion in its wake.

Today we know that this "Carrington Event", the most violent geomagnetic storm in recorded history, began with a particularly powerful solar flare – a vast burst of energy unleashed when the tangled magnetic field around a group of sunspots found a way to short-circuit and reconnect at a lower level. The liberated energy

---

* For a great account of this pivotal event in our understanding of the Sun, check out Stuart Clark's *The Sun Kings*.

heated nearby gas to millions of degrees and triggered a coronal mass ejection (CME) – an expanding bubble of superheated gas, vomited out of the solar atmosphere at millions of kilometres per hour and still imbued with the shredded remnants of its original magnetic field.

The solar cycle, it seems, affects not just sunspots but also the frequency and strength of flares and CMEs. A modern-day Carrington Event would have the potential to wreak havoc, disrupting the technology and communication networks that underpin modern society. And in the long run, such an event is inevitable.

* * *

The Sun's proximity to Earth also allows it to act as a laboratory for ideas about the internal structure of stars and the energy sources that underlie it. One of the first people to think seriously about this stuff was Arthur Stanley Eddington, a brilliant astronomer and physicist who made his reputation through his championing of Einstein's theory of general relativity during and shortly after World War I.

In his 1926 book, *The Internal Constitution of Stars*, Eddington suggested for the first time that the Sun's energy was generated in its central core (where temperatures and pressures are at their highest), rather than at its surface. If this was the case, then in order for the Sun – or any other star – to remain stable, each layer must find a balance between the inward pull of gravity and the pressure exerted by escaping radiation from below.

Eddington's model of this "hydrostatic equilibrium" suggested that stars like the Sun had a three-layered internal structure. After being produced in the core (the central 25%

of the Sun's radius), energy escapes in the form of high-energy radiation (mostly X-rays and gamma rays with much shorter wavelengths than visible light) through a "radiative zone" where the Sun's interior is technically transparent but in fact forms a near-impenetrable fog due to the sheer density of matter. The radiation takes around 170,000 years to fight its way across this zone, with each packet of energy going through about ten million billion billion collisions that slowly rob it of energy and transform gamma rays to X-rays and ultraviolet light.

Just over two-thirds of the way towards the surface, the falling temperature and pressure mean that matter no longer deflects radiation, but absorbs it instead. The base of this opaque zone soaks up the heat from below, causing it to expand and float up towards the surface, jostling its way through overlying material that is cooler and denser, and creating churning cells where heat is transported by convection (the bulk movement of matter) rather than radiation. The photosphere marks the top of this "convection zone", where the density falls enough for the Sun's gases to become transparent once again, and energy can escape in the form of visible light*.

But what was actually generating all that energy in the core? Here, as we'll see, Eddington made a remarkably lucky guess.

* * *

The question of what makes the Sun (and by extension other stars) shine only became troublesome after nineteenth-century

---

* The fundamentals of this model have been confirmed by the remarkable science of helioseismology, which measures Doppler shifts, created as different parts of the photosphere move towards and away from Earth, in order to detect seismic waves passing through the Sun's internal layers.

breakthroughs in geology made it clear that the Earth was much older than the 6,000-odd years suggested in the Bible. If the Earth had been around for millions of years, then clearly the Sun must have too.

In the late nineteenth century, the favoured explanation was gravitational contraction – if the Sun was indeed a fluid, then perhaps it was being heated by slow contraction under its own gravity? This idea was first floated by the polymath German scientist Hermann von Helmholtz in 1854, but really took off when respected physicist William Thomson (later Lord Kelvin, famous for defining absolute zero amongst other things), crunched the numbers in 1870 and found that a Sun powered in this way could shine for perhaps 20 million years[*].

At first, this seemed long enough to accommodate the known or likely span of Earth's history, until further breakthroughs in the early twentieth century extended it yet further. The discovery of natural radioactivity and the invention of radiometric dating[†] rapidly telescoped the age of the Earth to several *billion* years – and there was no known power source that could fuel the Sun for that long.

Fortunately, if radioactivity revealed the problem, then it would eventually also offer the solution. As physicists got to

---

[*] The "Kelvin-Helmholtz" mechanism might not be a long-term solution for keeping stars shining, but it's still important to stellar newborns (and gassy planets), so in brief: heat escaping from the surface leads to a drop in the star's internal pressure so the inner layers shrink in volume. This raises the pressure and temperature in the core, and heat escapes outwards.

[†] In its simplest form, if you know the rate at which an unstable radioactive element decays into something else, and can be sure that the "something else" wouldn't otherwise be present in your sample, then measuring the proportions of the two substances reveals just how long the original isotope has had to decay since the rock formed.

grips with the structure of atoms and the tiny particles inside them, they eventually discovered that, just as the central nuclei of heavy atoms could split apart through radioactive decay and release energy in "fission" reactions, so the nuclei of lightweight atoms could be joined together in "fusion" reactions. The products of fusion had slightly less mass than the pieces pushed together to form them, and that tiny difference in mass could be converted directly into energy using Albert Einstein's famous equation $E = mc^{2*}$.

Fusion was the solution proposed by Eddington in 1926 – and by way of a demonstration, he discussed the 0.8% difference between the mass of a helium nucleus and the mass of the four protons (the central cores of hydrogen atoms) that would be needed to create it in a fusion process. At the time it just seemed like a neat example, since despite the huge advances of the preceding decades, no one really had any idea what the stars were made of.

Well, perhaps one woman did. Cecilia Payne was a Buckinghamshire-born Cambridge science graduate, whose interest in astronomy had been ignited when she saw Eddington lecture in 1919. Now studying for her doctorate at Harvard College Observatory, she was using the vast Henry Draper Catalogue (see Aldebaran) of stellar spectra to investigate *how* the different spectra arose.

The nagging problem was that, taken at face value, the huge variety of spectra suggested an equally great variation in the makeup of the stars. Astronomers had become proficient in linking different absorption lines to different atoms, but they were

---

* With "c" standing for the speed of light.

still puzzled by their implications: if the strength and number of spectral lines truly represented the proportions of the elements they were associated with, then some stars must be dominated by carbon, others by oxygen, and so on. This seemed rather unlikely; why woudn't stars all be made from the same basic material?

In search of alternative explanations, Payne turned to the recently discovered "ionisation equation" of Indian astrophysicist Meghnad Saha – a way of describing the conditions in stellar atmospheres at different temperatures, and in particular the way that subatomic electron particles were gradually stripped away from the atoms of gas to create a variety of electrically charged particles called ions. The more electrons an atom had in its easily removed outer layer, the greater variety of "ionisation states" it might have – if temperatures were sufficiently high to strip those electrons away.

When Payne considered what this might mean for spectral lines, everything fell into place. The variety of spectra turned out to reflect the range of ionization states present among more or less the same basic mix of elements – the differences were mostly down to the fact that hotter stars have more states "switched on" and accessible to form spectral lines. Payne was soon able to show for the first time that the stars contained elements such as silicon, carbon and oxygen in similar ratios to those found on Earth. But these relatively heavy elements were only a small proportion of the atmosphere – the rest, the ionisation equation suggested, was dominated by helium and, especially, hydrogen.

By 1925, Payne had her thesis ready, but when one reviewer – none other than Henry Norris Russell, the man who put the "R" in the H-R diagram – advised her she'd be wise to take out the nonsense about hydrogen and helium, she reluctantly conceded[2]. Her work still made a huge impact, but its most important aspect lay on the cutting-room floor, even as Eddington was putting together his own thoughts. It is ironic, then, that it was Russell who ultimately revived the idea of hydrogen and helium as the dominant elements of stars, in a 1929 paper analysing the Sun's spectrum using a different technique, yet reached pretty much the same conclusions[*][3].

All this paved the way for the physicists who would finally come up with a plausible fusion mechanism in the 1930s. Overcoming the mutual repulsion between particles with identical electrical charge such as protons was clearly a challenge, but in 1928 Russian physicist George Gamow used the strange new science of quantum mechanics to show that it could occasionally be done amid the high temperatures and pressures of the Sun's core. Within five years, Gamow had defected to the US, where with his German colleague Carl Friedrich von Weizsäcker he developed the idea of a "proton-proton chain", in which collisions between the nuclei of hydrogen atoms gradually built up to form a nucleus of helium – the very example that Eddington had provided.

There were still a few wrinkles to iron out, however. In particular, Gamow and Weizsäcker's theory had a tough job explaining how some particularly unstable particle clusters avoided bursting into pieces well before they could collide with others. It took an outsider, in the form of Hans Bethe, a German Jewish émigré to the States, to find a more stable path for building helium. By 1939, for the first time, Bethe and

---

* Russell did, to his credit, point out Payne's earlier work on the topic.

his colleague Charles Critchfield had explained how a blend of different fusion reactions could create small amounts of elements heavier than helium – a process called nucleosynthesis that continues in the Sun and all other main-sequence stars today[4].

PROTON-PROTON CHAIN

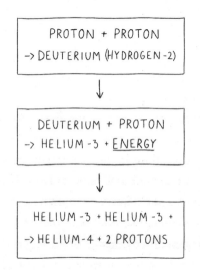

7 – THE TRAPEZIUM AND OTHER WONDERS

Nebulae and the origins of stars

※

Okay, so apparently stars are enormous balls of gas – floating on their own in vast empty gulfs of space and shining thanks to nuclear reactions in their cores. Isn't that a bit unlikely? How did they get there in the first place? How do binaries like 61 Cygni and multiples like Mizar end up in perpetual orbit around each other while loners like our Sun finish up light years away from their nearest neighbours?

The answers to these questions turn out to be linked to some of the most beautiful objects in the night sky – emission nebulae (*nebula* is simply Latin for cloud). These nebulae, defined by the fact that they emit their own light, rather than just reflecting it from elsewhere, are one of several different types of gas and dust cloud that permeate the space between the stars, and can themselves become the sites of starbirth. So in this chapter, for a change, we don't have to worry about deciphering the shifts and blinks of a single point of light or even the waltz of a dancing pair: instead we can lose ourselves among the swirls and ripples of vast and delicate cosmic clouds.

The great constellation of Orion is one of the most obvious in the entire sky, a square-shouldered, tunic-clad giant with brilliant stars, Betelgeuse and Rigel, at his left shoulder and right knee, and a cinched waist marked by a belt of three stars (see page 196 for a chart). For Northern Hemisphere skywatchers this celestial hunter vaults across the southern evening sky between November and April (early risers can spot him from August), while in the Southern Hemisphere he performs an elegant backflip above the northern horizon in the same months.

Orion is the most fully formed constellations, with stars that mark both shoulders and knees, as well as two raised arms and the club and shield they carry. His feet are missing (let's charitably say they're hidden behind the hare Lepus), and the head is marked only by a triangle of faint stars, but the rest is instantly recognisable. And hanging from his belt, a chain of stars marks Orion's sword.

Here, most naked-eye observers can make out three stars, but it's obvious at a glance that there's something different about them, and binoculars will quickly reveal what's up. The upper and lower stars of the sword resolve into small groups of stars (a pair at the bottom and a more complex group at the top), while the middle one – well, the middle one is very peculiar indeed.

Depending on how dark your sky is and what sort of device you're looking with, the middle of Orion's sword can look like anything from an indistinct triangular smudge of light to a beautiful rose, with three or four jewel-like stars at its heart*. This

---

* Higher magnification will help you separate the individual stars, but as you zoom in you'll lose light from the surrounding nebula. So, unless you're a double-star nerd, it's better to settle for a low magnification and a nice wide field of view that will allow your eye to soak up as much light as possible.

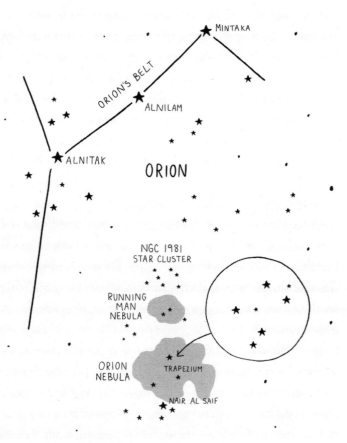

is the Trapezium, a group of four newborn stars at the heart of a broader star cluster. The glow that surrounds it is the Orion Nebula (sometimes just called the Great Nebula) – the largest and brightest star-forming region in Earth's skies.

The nebula's glow is equivalent to that of a magnitude 4.0 star, so it's curious that stargazers didn't pay it much attention until the invention of the telescope. Even Galileo seems to have passed it by when looking at the region, and thus the first official record of its existence comes from one of his contemporaries and correspondents, French natural scientist Nicolas-Claude Fabri de Peiresc, who spotted it on the night of 26 November 1610. An absence from early star catalogues such as those of Ptolemy and the great Islamic astronomer Al-Sufi has led to some speculation that the nebula brightened up considerably around 400 years ago. We'll see in the next chapter how that sort of thing isn't entirely out of the question – but in this case it's more likely that Galileo's telescope simply wasn't good enough to catch the surrounding haze.

What Galileo did spot, however, were the individual stars at the centre of the nebula. Previously, the central star of the sword had been catalogued using the simple Greek lettering system as Theta Orionis, but through a small scope or good binoculars it resolves itself into a rough quadrilateral shape, looking a bit like a pyramid with the top lopped off *. Galileo apparently spotted three of the stars in this trapezium, and Johann Baptist Cysat, a Swiss Jesuit priest and astronomer, added the fourth when he published his description of the nebula in 1619. As telescopes improved, more and more stars were inevitably spotted, but the identification with a trapezium-shape stuck and eventually became a byword for the entire cluster.

* * *

The true nature of nebulae like the one in Orion's sword would remain a puzzle for two and a half centuries, even as telescopes improved and many more of them were discovered.

---

* In technical terms, a trapezium is generally understood to be a quadrilateral with one pair of parallel sides – the base and top surface of our lopped-off pyramid – but in the US, for obscure reasons, the word is applied to any non-rectangular four-sided shape.

While many fuzzy objects in the sky ultimately revealed them-selves to be compact star clusters, others remained resolutely hazy.

That didn't stop speculation, however, and several far-sighted thinkers wondered whether the nebulae might be linked to the birth of stars. First off the blocks was Emanuel Swedenborg, a remarkable Swede who spent the first half of his career as an extraordinarily prolific scientist, theorist and inventor, before a turn to the mystical saw him lay the foundations of a Christian sect that remains active to this day.

Swedenborg prided himself on his habit of throwing ideas at the wall to see what stuck, and one of these was his "nebular hypothesis" of 1734 – the idea that the Sun and planets condensed out of material that was originally a gaseous nebula. Although the theory of matter underlying his ideas was wildly off-beam by modern standards, he nevertheless deserves the credit for the original idea that others later ran with[1].

One of these others was none other than Immanuel Kant, German philosopher and butt of *Monty Python* jokes. Long before his hugely influential works on metaphysics, Kant took an interest in what was then called natural philosophy, laying out a "theory of the heavens" in 1754 that developed Swedenborg's ideas in line with Newtonian physics. For astronomers (who as a rule are a practical folk and not given to metaphysics), Kant is best known for his speculations that some nebulae might be the locations of present-day starbirth, while others might in fact be distant galaxies – vast star systems far beyond the Milky Way that most of his contemporaries viewed as the entirety of the Universe.

During the 1790s, the idea of stars forming from nebulae won further influential backing from both French theoretician

Pierre-Simon Laplace (who put it on a sound mathematical footing) and the Hanover-born "Prince of Astronomers" William Herschel.

Revisiting the list of bothersome objects catalogued by French comet hunter Charles Messier, Herschel identified many of them as compact clusters of stars, but was equally convinced that some contained luminous clouds of gas he called "bright fluids." Because these clouds often had stars and clusters embedded within them, Herschel pinpointed them as potential stellar nurseries, with "Messier 42" – the vast and intricate Orion Nebula with the Trapezium at its centre – the most obvious example.

Over the next two decades – with the backing of King George III's royal chequebook – Herschel continued to observe nebulae and star clusters through ever-larger telescopes, until by 1814 he was convinced he had traced every stage in the birth of stars through condensations of matter in the bright fluids[2]. From a modern point of view, the major problem with his theory is that much of it was the wrong way round. Herschel believed that stars were born individually and then gradually accumulated into clusters, so that widely dispersed clusters would be younger and the denser ones older[*].

It took another 50 years before further advances moved the question on. In 1864 spectroscopy maven William Huggins (fresh from his triumphant identification of elements in the

---

[*] In 1888, Danish-Irish astronomer J.L.E. Dreyer recognised an important difference that might have saved Herschel's confusion. His New General Catalogue – a vast expansion of the Messier list of non-stellar objects – distinguished between "open clusters" containing dozens or hundreds of stars, and far more densely packed balls of stars we now call globular clusters (see Omega Centauri). Only the open clusters are associated with nebulae and sites of star formation.

atmospheres of Aldebaran and other stars), decided to take a look at the nature of light from nebulae.

It being summer, Huggins didn't bother waiting for the obvious target of the Orion Nebula to reach its evening apparition, but instead pointed his telescope at another curious object in London's circumpolar sky, known as the Cat's Eye. This bright "planetary nebula" – one of many broadly ring-shaped clouds whose true nature would only become apparent in the 1950s (see Sirius B) – immediately displayed a spectrum that was unlike any star. Instead of showing dark lines against a bright background, caused by elements catching and absorbing starlight in transit, it revealed a few bright lines on a mostly dark background – similar to the emission spectra from heated vapours with which Huggins had been comparing the light of stars.

As soon as Orion rose back into view, Huggins wasted no time in showing that the Great Nebula was also a cloud of glowing gas – albeit one with a very different structure to the Cat's Eye. Rather than sticking his neck out to support the nebular hypothesis, Huggins chose at the time to class all gaseous nebulae as a completely different kind of object, independent of the stars.[3]

The photographic revolution, however, would soon leave the link between starbirth and emission nebulae beyond doubt. The first photograph of the Orion Nebula, captured by New York pioneer Henry Draper in 1880 using a 50-minute exposure, was little more than a blurry proof-of-concept, but just three years later, Newcastle-born Andrew Ainslie Common[*] captured a

---

[*] A Geordie bear of a man who made his fortune from the Victorian boom in sanitary engineering, and then spent it on a series of increasingly large and ambitious telescopes.

groundbreaking image by combining several long exposures to dazzling effect.

Common's photograph (above), for which he won the Royal Astronomical Society's Gold Medal, wasn't just a pretty picture – it also showed for the first time that photographs could capture detail and faint stars beyond the perception of the human eye. The Trapezium itself was lost in the blaze of light at the heart of the nebula, but the three-dimensional structure and the extensive field of stars around it were worth the trade-off.

Photography would soon reveal other types of matter between and around the stars, all of which would play their part in the story. In the 1890s, E.E. Barnard at the University of Chicago's Yerkes Observatory, began a survey of the dark gaps in the Milky Way that Herschel had referred to as "holes in the heavens". Working with Max Wolf of the University of Heidelberg, he eventually concluded that these apparent holes were in fact dark nebulae – clouds of opaque dust blocking the light from more distant stars beyond. Then in 1912, Vesto Slipher, an Indiana-born astronomer working at the remarkable Lowell Observatory in Flagstaff, Arizona (of which more later) concluded that the glowing blue nebula around Alcyone's sisters Merope, in the Pleiades cluster, was caused not by the emission of visible light but by simple reflection of starlight from Merope itself[4]. Looking at modern photos of the Orion Nebula today, it's easy to see where these dark nebulae and reflection nebulae also play a part in shaping the overall structure.

But despite increasing certainty that emission nebulae, with the exception of the curious "planetary" type, were the sites of starbirth, the precise sequence of events remained frustratingly hard to pin down with no accurate means of working out how old the stars in different clusters and nebulae actually were.

The key turned out to lie in a concept we've already encountered. Remember Ejnar Hertzsprung's comparison of the H-R diagrams (see Alcyone) for the Pleiades, Praesepe and the Hyades? And how the Pleiades has stars that go all the way to the bright, blue top end of the main sequence while the others are "missing" these stars, but have brilliant red and orange ones in their place? As astronomers began to get to grips with the processes of stellar ageing in the mid-twentieth century, they came to realise that the red and orange stars are a later phase of a life cycle that runs at different speeds depending on how brightly the star shines while it is sitting on the main sequence. The brightest, hottest stars evolve fastest and soon start to move off the main sequence, stripping away the upper end of the sequence within a particular cluster as they transform into luminous orange and red giants.

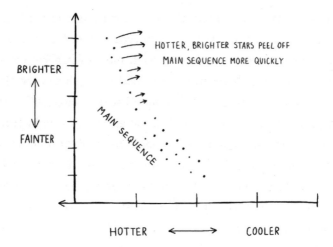

So the degree of erosion in a cluster's main sequence (and the number of red and orange stars) is a handy tool for measuring its age, at least in comparison with other clusters. This realisation confirmed beyond doubt that the youngest, brightest and bluest clusters (often still embedded in nebulosity) were the most densely packed, while older ones had a tendency to drift apart. In 1947, an Armenian astronomer called Viktor Ambartsumian discovered the next stage in this evolution. As clusters disperse, they give rise to "OB associations", groups of moderately hot and bright stars (belonging to spectral classes "O" and "B") scattered across a large region, whose proper motions can all be traced back to the shared spot where they were born.

All of this means that the Trapezium and other stars at the heart of the Orion Nebula are very young – no red giants here, and plenty of material to fuel continued star formation. The Trapezium stars themselves are all junior heavyweights, each with an estimated mass of more than 15 suns. The brightest of all, catalogued as Theta Orionis C, is the southernmost of the main quadrilateral, with an individual magnitude of 5.13, and a blue tint flagged up by its spectral type of O6. It's actually a binary system in its own right, though most of its light comes from a single star. Measuring its orbital motions has revealed that this monster weighs in at about 33 solar masses. With the distance to the Trapezium estimated at about 1,330 light years, this means the star pumps out more than 200,000 times more energy than the Sun. The only reason it's not brighter in Earth's skies is because much of this energy is emitted not as visible light, but as invisible ultraviolet radiating off a photosphere heated to an incandescent 39,000°C.

In fact, all the Trapezium stars pump out vast amounts of ultraviolet, and this turns out to be the beauty secret of the spectacular cocoon that surrounds them. The Orion Nebula, like all emission nebula, is shining through the same mechanism as the striplights over your kitchen counter.

Fluorescence is a natural phenomenon that has been known since the early 1600s – emission of light from an otherwise unremarkable material that has itself been illuminated, with the glow often continuing once the original light source is removed. The first person to explain it properly was French experimentalist Edmond Becquerel – father of the radioactivity pioneer Henri, and probably best known today for building the first solar cell.

In 1842, Becquerel discovered that calcium sulfate (a common chemical found as the mineral gypsum, and refined into plaster of Paris) fluoresces in visible light when exposed to ultraviolet. He noted that the wavelength of the incoming UV was shorter than that of the light waves given off, although it was Anglo-Irish physicist George Gabriel Stokes who, a decade later, suggested this was an immutable law of fluorescence.

Today it's clear that emission nebulae across the sky glow due to excitation. Their gases are energised by ultraviolet rays from newborn monster stars in the vicinity, and as individual atoms shed this excess energy, they do so in stages, shedding less energy at each step and emitting longer wavelengths of visible light.

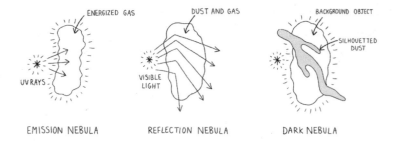

EMISSION NEBULA   REFLECTION NEBULA   DARK NEBULA

So, stars are born from condensing clouds of gas, and as they emerge, their radiation transforms these clouds into emission nebulae. Initially they form dense clusters with stars all the way along the main sequence (that crowded diagonal band that dominates the H-R diagram), but as they age, their brightest stars transform into red giants and eventually disappear completely. As the supply of ultraviolet radiation from bright, hot stars dwindles, the nebulae are robbed of their power supply and begin to fade, while clusters themselves drift apart and finally lose their integrity altogether.

We'll take up the story of star formation at the level of individual stars in the next chapter where we visit the remarkable T Tauri, but one last question to ask at this point is just how the collapse process somehow goes into reverse. Why do the newly formed stars eventually disperse across space instead of orbiting closer and closer until they eventually merge together into some monstrous supersun?

The answer to that is intriguing, and sends us back to another star in Orion. Look just below the Orion nebula and you'll find Iota Orionis, the brightest star in Orion's "sword", otherwise known as Na'ir al Saif. This blue giant of magnitude 2.77 is actually a spectroscopic binary with a close companion in an orbit that lasts a mere 29 days. They share a near neighbour of magnitude 7.7 (visible with binoculars), locked in an orbital waltz that seems them spinning around each other once every 75,000 years.

Na'ir al Saif seems to mark the site of a marital bust-up on a cosmic scale – it sits at an intersection point where the proper motion of two distant heavyweight stars, both now some way from Orion, can be traced back over 2.5 million years. Computer simulations suggest that these two exiled stars (known as Mu Columbae and AE Aurigae) both started out as members of monster binary systems that had a close encounter in the heart of the Orion Nebula, near the current Trapezium[5]. During this brief but fateful encounter, one star in each system found itself lured away by the seductive force of the Iota Orionis close binary, while the rejected former partners were cut loose, fleeing across the sky in opposite directions at a relative speed of 200 kilometres per second.

Although this episode of stellar infidelity is particularly notable for the size of the stars involved, most astronomers think that a similar mechanism – close encounters and partner-swapping between stars – is ultimately responsible for the slow disintegration that most star clusters (with a few notable exceptions) undergo.

8 – T Tauri

The nitty gritty of starbirth

⨯

We know from studies of regions like the Great Orion Nebula that stars are born in huge clusters, as vast clouds of interstellar gas and dust coalesce. But how do individual stars actually form, and how do they behave in their early years? Not far from Orion's Trapezium cluster lies a star that played a pivotal role in revealing this story – although it's one of the more testing objects on our tour of the sky.

T Tauri* takes us on another visit to the bull constellation of Taurus, one of the most obvious star patterns in the sky. You'll need a small telescope or very good binoculars to have much hope of spotting this influential infant star, but the best route is to start with the V-shaped star cluster of the Hyades forming the bull's face.

---

* Why T? Well, adding to the soup of Bayer letters (Greek letters to indicate a star's brightness within its constellation) and Flamsteed numbers (which map the locations of a constellation's mid-ranking stars from west to east), astronomers designate otherwise un-catalogued variable stars with an unholy mess of letters and numbers. The only ones that need concern us, fortunately, are the capital letters R through Z, which are used to designate the brightest variables in a particular constellation.

If you think of the "V" as having two arms, with the end of the southern arm marked by the brilliant Aldebaran, then we're interested in the tip of the upper arm – a fainter yellow star called Epsilon Tauri. Find this using binoculars and then look a little to its west, where you'll find a wedge-shaped triangle of fainter stars. This wedge points in a southeasterly direction, but if you now move in the opposite direction for about three times its length, you *should* come to a pair of fainter stars in roughly the same orientation as the wide end of the wedge. The northernmost and fainter star in the pair is T Tauri.

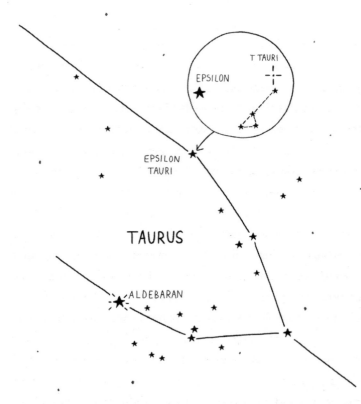

We say *should*, because T Tauri is a capricious little devil. As we saw way back at the start of our story with Polaris, many stars in the sky actually vary in brightness to some extent, but T Tauri is the first of several extreme cases we'll be meeting. Its light output fluctuates unpredictably between around borderline binocular visibility at 9.3, and a lowly magnitude 14. In fact, you may find that you're just as likely to spot a nearby smudge of light – the nebula that led to the discovery of T Tauri itself.

T Tauri and its accompanying nebula were both discovered in 1852 by John Russell Hind, a Nottinghamshire lacemaker's son who had joined the Royal Observatory at the age of 17 and was now director of a private observatory set up by the wealthy winemaker and merchant George Bishop near his home in Regent's Park. Hind had already established something of a reputation through his discovery of several asteroids and a remarkable stellar explosion or nova (for more on these, see RS Ophiuchi). The report of his new discovery focused first on the nebula, and only mentioned in passing that the tenth-magnitude yellow star next to it seemed to have been missed off previous catalogues.[1]

The unpredictable nature of star and nebula became apparent over the following years. By the late 1850s, it was clear that T Tauri's brightness was declining, and a few years later both star and nebula vanished. This sort of behavior was not so surprising for a star, because, by this time, many other so-called variable stars had been discovered (we'll be looking at two of the most famous shortly) – but a well-attested variable nebula was new and unusual.

Several of the stargazers who went looking in the area over the next few decades got thoroughly confused. In 1868, German astronomer Otto von Struve (son of parallax-hunter Friedrich) reported that T Tauri had brightened again and was accompanied by a small glowing nebula – though in a different place to that seen by Hind*. In 1890, Sherburne Wesley Burnham, a supremely talented amateur astronomer taking a sabbatical from his profession as a court reporter to work at the University of California's Lick Observatory, found that T Tauri *itself* was nebulous, appearing as fuzzy, elongated oval under high magnifications.[2] Owing to a misunderstanding, Burnham thought he'd recovered Hind's nebula. It was only in 1895 that his partner for the original observations, Edward Emerson Barnard, returned to T Tauri and finally made sense of it all. He discovered that the nebulous appearance of the star had disappeared, and there was once again a nearby gas cloud in the exact space recorded by Hind.[3]

So what the heck is going on? The most obvious way a nebula can vary is if it's powered in some way by a nearby star – for instance either through the simple reflection of starlight, or by the fluorescence mechanism we saw lighting up gas around the Trapezium.

Unfortunately, the hope of a simple explanation was dealt a blow by the fact that the fluctuations of Hind's Variable Nebula (as it is now known) don't always match with T Tauri. That's probably because there are other, unseen players in the game. In 1982, astronomers attempting to resolve the details of the region around T Tauri with the VLA radio telescope (a giant's train-set of movable dishes in the desert near Socorro, New Mexico) found that much of the star's infrared emission was coming from a hitherto unseen companion[4]. The two stars are separated by

---

* This particular nebula soon disappeared, never to return – hence the magnificently intriguing name of Struve's Lost Nebula, listed a few years later in J.L.E. Dreyer New General Catalogue as NGC 1554.

about three times the distance between the Sun and Neptune (not a lot in cosmic terms), but the companion is invisible to optical astronomers not only due to obscuring gas and dust, but also because it is significantly cooler with a surface that is barely red-hot. Further studies have shown that the new "T Tauri S" star (so-called because it's the southern element of the pair) is itself a close binary, making for a triple star system overall.

All three elements of T Tauri lie within a large, dark cloud that spans much of the Milky Way between Taurus and the neighbouring constellation Auriga (the Charioteer) to its north. Since the discovery of the original star, many more fluctuating stars with similar properties have been found, scattered across the sky but often in proximity to dark, obscuring clouds that are also packed with unseen cool objects emitting infrared rays. Although varied in colour, from white and yellow to orange and red, they all share similar properties including unpredictable brightness fluctuations, strong emissions of both high-energy X-rays and low-energy radio waves, and the presence of nearby nebulae. What's more, their spectra routinely display absorption lines from the lightweight element lithium (the Universe's third lightest and most common element after hydrogen and helium). In 1945, American astronomer Alfred Harrison Joy suggested that stars with these properties could be treated as a distinctive class – the T Tauri variables.[5]

Together, the T Tauris seem to represent a broad range of different stars that are just beginning to shine. The presence of lithium is a particularly important clue here. This lightweight element is plentiful in the star-forming gas but easily destroyed by high temperatures once stars get going – if a star has lithium in its atmosphere, then it can't have been shining for long.

So how exactly *do* stars get started on their long journey through life? The earliest stage in the formation of an individual star system (whether single, binary or multiple) seems to be a dark, opaque clump known as a Bok Globule, named after the Dutch-American astronomer Bart Bok who first drew attention to them in 1947[6]. With diameters of up to a light year across, Bok guessed that these might be the cocoons in which stars were forming, but it took a long time for him to be proven right. By the late 1960s, astronomers had noticed the huge numbers of Bok globules in and around nebulae such as the Orion Nebula, and even noted with interest that their behaviour suggested each contained about a star or two's worth of material. However, the crucial evidence only came after the successful mission of IRAS, the InfraRed Astronomical Satellite[*] in 1983.

Even then, sifting through IRAS's data took a long time, and it wasn't until 1990 that astronomers were able to confirm the link between the opaque globules and many of the sky's infrared hot spots.[7] Interest in the minutiae of star formation really kicked off a few years later when the Hubble Space Telescope, fresh from the repair mission that fixed its faulty eyesight, turned its gaze on one of the sky's largest emission regions, the Eagle Nebula in the constellation of Serpens. The word "iconic" is bandied around a lot these days, but among space nerds, Hubble's image of the so-called "Pillars of Creation" is up there with anything the Kardashians have to offer.[8] The heart of the nebula was revealed as an alien landscape dominated by teetering towers of

---

[*] Not the most imaginative name, but this joint NASA/UK/Netherlands space observatory was the first of its type, getting above Earth's muggy atmosphere and using a super-cooled telescope to spot sources of heat radiation across the sky during its landmark 300-day mission.

opaque gas and dust where the actual business of star formation gets going, from which probing trunks and tendrils emerge into the surrounding open space.

The key to understanding what was going on in this Gigeresque fantasy turned out to lie in the eerie glow around the top of the pillars. This is a haze produced as the gas and dust inside are blasted by UV radiation from the brilliant stars that have already emerged from the pillars and begun to shine in the region above them (as a general rule, the most massive stars form the fastest, and it is their radiation that switches on the emission from the rest of the nebula). The glow highlights a process called photoevaporation, in which the UV radiation first splits the gas atoms and molecules apart into electrically charged ions, and then blows them away towards the edges of an expanding cavity within the nebula. Relatively dense regions where other stars have begun to form can better resist this effect, but as their surroundings are beaten back, they are untimely ripped from the cocooning pillars, remaining attached to them only by a tenuous umbilical that falls broadly within their protective shadow. Researchers Jeff Hester and Paul Scowen, with an eye for a good acronym, named these emerging nuclei of star formation "evaporating gaseous globules". Each EGG contains the nucleus of either a single star, or a more complex binary or multiple whose members will be born locked in orbit around each other.

And now we come to one of nature's greatest conjuring tricks – the bit where a shapeless cloud of slowly rotating gas and dust is transformed into one or more newborn stars surrounded by a flattened disc of left-overs ready to form a solar system. As with many tricks, the vital moment is frustratingly hidden from our sight – not in this case by a curtain-wielding assistant, but by the opacity of dust in the globule itself. Fortunately, however, some basic physics can give us a pretty shrewd idea of what we're missing.

The process hinges on a fairly mundane rule called the "conservation of angular momentum". If you've ever watched figure skating, you've seen this in action when a pirouetting skater pulls their arms inwards and transforms themselves into a fast-moving blur. Angular momentum is just a measure of the momentum of a spinning or orbiting object, and it depends not just on mass and speed of movement (like normal straight-line momentum) but also on distance from the central axis of rotation.

The important thing from our point of view is that if a self-contained "system" (physics-speak for anything from a spinning skater to an EGG of star-forming gas) is free from outside forces, its angular momentum can't change. If you concentrate matter towards the middle, it has to spin faster and faster just to keep the sums balanced.

So just like that skater, the star-forming globule gradually pulls its arms in through the inexorable effects of gravity, drawing mass towards its centre and spinning ever faster. At the same time, collisions between gas and dust clouds moving in slightly different directions start to jostle everything into a flattened circular disc*.

As material falls in towards the centre of the disc, it carries with it the energy associated with its motion, releasing it as heat

---

* To take another ice-skating metaphor, imagine a crowd of skaters wobbling their way around a circular ice rink – they'll soon settle into concentric circles to avoid collisions, and anyone who gets ideas about cutting across the stream of traffic is likely to find themselves pushed back into line in no uncertain terms. The same thing goes for gas clouds orbiting stars, except it works in the up-down dimension as well.

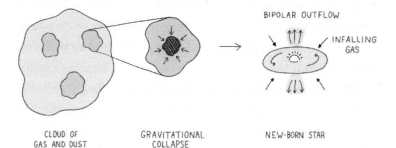

CLOUD OF   GRAVITATIONAL   NEW-BORN STAR
GAS AND DUST   COLLAPSE

that makes all the atoms in the cloud jiggle around a little bit faster. Normally this would tend to make the atoms fly apart, but in this case, the rapidly increasing pull of gravity wins out, and the middle of the cloud just gets hotter and denser. At a somewhat vaguely defined point, the dense gas cloud crosses the threshold to become a "protostar" – a compact object that generates enough energy to make its presence known at infrared wavelengths, but which is still accumulating gas from the surrounding globule.

A star like the Sun may spend about half a million years as a protostar, with temperatures and pressures steadily increasing in its core. As it gathers more and more material, the gravitational contraction making it shine actually pumps out *more* energy than a "normal" star of similar mass*. This powers a strong stellar wind of particles that blow away from the surface, gradually bringing the steady accumulation of infalling gas to a halt and eventually driving it back. The cocoon begins to shred and fragment, even as the energy emitted from the collapsing protostar continues to increase and the escaping radiation clambers from the relatively

---

* Remember when nineteenth-century greats Helmholtz and Kelvin tried to invoke gravitational contraction as a potential power source for the Sun? The problem isn't *how much* energy the process can produce, just *how long* it can last for.

long-wavelength, low-energy infrared towards the visible part of the spectrum.

The shift from steady accumulation of material to its active rejection often triggers spectacular changes to the environment around the protostar, which were studied by Alfred Joy's friend and former student, George Herbig, in the 1960s. Herbig realised that while material falling onto the protostar would cause it to spin at increasing speed (that ice-skater principle again), the build-up of angular momentum would eventually overcome this. New material falling onto the star's equator would be moving fast enough to achieve escape velocity, allowing it to overcome the attraction of gravity and be ejected back into space. Because the star's equator is still surrounded by a thick doughnut of dense, relatively cold gas and dust, this hot, fast-moving material is channeled out in two "bipolar" jets from the star's poles. Where the jets slam into other matter in the wider nebula or interstellar space, they energise it in a similar way to UV radiation. The result is a pair of glowing clouds emerging on either side of the still-cocooned infant star, known as Herbig-Haro (HH) objects (named after George and Mexican astronomer Guillermo Haro, who identified these structures independently). Bipolar jets and HH objects shine for just a few thousand years before fading, but they are spectacular objects while they last.

* * *

The exact nature of the star that eventually emerges blinking into the cold light of interstellar space depends, above all, on its mass. T Tauri stars tend to have less than three suns' worth of material in them. They start out bloated in size and more luminous than adult stars of a similar mass, but slowly shrink and fade over the course of about 100 million years as their gravitational power source is used up. Pictured on a H-R diagram (see Alcyone), they're effectively dropping down a vertical line called the Hayashi track*, maintaining the same colour despite losing their initial brilliance.

Even while the star is fading, though, contraction is still causing its central regions to grow ever denser and hotter. The combination of pressure and temperature break gas molecules into their component atoms, and strip away electrons that orbit in the outer layers of the atoms to leave exposed atomic nuclei. This paves the way for low-level nuclear reactions to kick off – not the full "proton-proton" chain that powers adult stars like the Sun, but instead a truncated version involving two naturally occurring variants or "isotopes" of hydrogen, called deuterium and tritium. With nuclei containing a single proton bound to one or two uncharged neutron particles, the two isotopes are effectively part-way along the chain already, with the most difficult job (pushing two lone protons together with enough force to overcome their mutual repulsion) already done[†].

A key event for heavier T Tauri stars (those with at least half the Sun's mass) occurs when they get hot enough to develop an internal radiative zone. At this point their luminosity becomes fixed, but a slow, inexorable contraction still continues, and the shrinking surface inevitably grows hotter. From this point on, the young star's evolution takes a sharp left turn across the

---

* Named after the brilliant Japanese astrophysicist Chushiro Hayashi, who figured a lot of this stuff out in the 1960s.

† Courtesy of the Big Bang, in which small but not insignificant amounts of deuterium and tritium formed alongside hydrogen and helium.

H-R diagram, following a horizontal track towards the diagonal main sequence.*

The T Tauri phase finally comes to an end when temperatures in the star's core get hot enough for full proton-proton fusion to begin (in the Sun's case, this took about 10 million years). The sudden increase in radiation flooding from the heart of the star finally stabilises its internal layers, and the resulting balance of luminosity, size and surface temperature determines the position at which the star will settle on the main sequence of the H-R diagram. It will remain here for the vast majority of its life – perhaps a hundred or a thousand times longer than this prelude. Over this vast lifetime, it will gradually be subject to similar processes and encounters to those we saw already working to break up the Trapezium cluster, breaking free from the fading nebula in which it formed, to eventually join the population of average, sunlike stars wending their way through our galaxy.

---

* As a rule, the more mass a young star has, the less time it spends going down the Hayashi track and the more it spends heading left and heating up. Roughly three solar masses mark the border between T Tauris and the so-called "Herbig Ae/Be stars", which show a much smaller drop in brightness and a much longer (and faster) drift to the left. By the time the most massive stars of all become visible, they are already in position at the brilliant blue end of the main sequence.

9 – Proxima Centauri

Dwarf stars that pack a
surprising punch

The closest star to our own solar system, just four and a quarter light years from Earth, is, sadly, not the most impressive object to look at. In fact, it can be a struggle to see at all, and it's insignificant enough to have been overlooked completely until the early twentieth century. But, nevertheless, Proxima Centauri (an unromantic name that means literally "the nearest star of the constellation Centaurus) is an inevitable stop on our tour of the heavens. Proxima not only offers an example of the cool, faint stars that make up the vast majority of the Milky Way and other galaxies, it also has some peculiar fascinations of it is own.

Proxima is actually the faintest member of a three-star system, vastly outshone by the close binary pair of Sun-like stars generally known as Alpha Centauri*. To track down Proxima, then, Alpha Cen is the best place to start – but here most Northern Hemisphere stargazers will be out of luck, as north of the tropics, these stars never make an appearance above the horizon. The constellation

---

* Sometimes referred to as Rigil Kentaurus, but given the dodgy spelling and risk of confusion with Orion's brilliant star Rigel, let's not even go there…

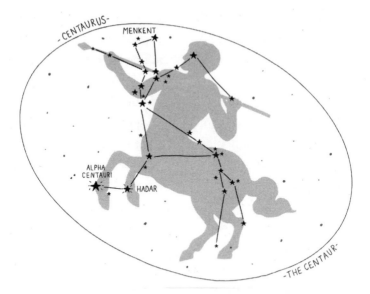

of Centaurus as a whole is fairly far south in the sky, and its two brightest stars, yellowish Alpha and bluish Beta (aka Hadar) are at its southern extremity (literally, since they mark the galloping Centaur's feet). However, if you're lucky enough to live south of about 25°N (so anywhere south of central Mexico, northern India or Taiwan), the two stars form a pretty unmistakable pairing as they pop up over the southern horizon. South of the equator, they're one of the most familiar sights in the sky, pointing the way to the Southern Cross tucked in ignominiously under the horse-man's belly*.

South of the Tropic of Capricorn, Alpha Centauri becomes a circumpolar star, and so spins around the dull region of the south celestial pole, dipping close to the southern horizon at times,

---

* You might be wondering at this point how a constellation that's barely visible from Greece gets named after a famous beast of Greek myth – we'll save the explanation for when we revisit this part of the sky to look at Omega Centauri.

but never actually setting. Around the equator, meanwhile, it's on show in evening skies between about May and September – and earlier in the year if you fancy pulling a late one or getting up before sunrise.

Take a look at Alpha through binoculars with decent magnification and you should be able to split its two stars most of the time (they orbit each other in a shade less than 80 years and are currently separating from a close alignment in 2015). A small telescope will do it for sure, and you'll need one anyway if you're serious about hunting for Proxima. Your best bet is to track south and west of Alpha until you come to a broad pair of stars about a degree away, creating the base of a narrow right-angled triangle, or wedge-shape, with Alpha at the tip. Now, try to imagine a mirror-image triangle, pointing towards the southwest: Proxima is more or less at its tip. If your observing equipment is up to scratch, then you should find a multitude of different stars to choose from, as this part of Centaurus also lies right in front of the Milky Way. Most of the stars in your field of view are much further away and intrinsically brighter than Proxima, but it's in there somewhere*, shining at a dim magnitude 11.1.

Proxima was discovered in 1915 by Robert Innes, Director of the Union Observatory in Johannesburg. Scots-born Innes was another astronomical amateur-made-good, having emigrated to Australia in his youth and made his fortune as a wine merchant before gaining a reputation as a shrewd observer and being invited to South Africa and into the academic fold.

---

* Sometimes as a backyard stargazer you just have to settle for knowing that a few of a star's photons are hitting the back of your eyeball, even if you can't pick out the source – but if you really want to give it a shot there are plenty of websites and apps out there to help.

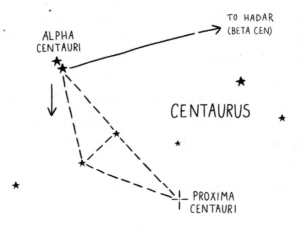

The discovery was a notable success for a new method of identifying stars on the move changing position from year to year against the more distant background with high "proper motion" across the sky. These were of interest because, as we saw with 61 Cygni, a star that moves rapidly from our point of view is, on balance, more likely to be nearby.

The technique involved a specially adapted viewfinder that could display one of two photographic plates to the examiner at the flick of a lever – not a million miles away from those machines they use to test your eyes at the opticians. By lining up two plates showing the same area of sky at different times (in this case, several years apart), it was simple to flip back and forth looking for objects that had moved*. Innes quickly realised that the new star he had discovered not only had a very high proper motion[1], but that its direction of motion matched with that of the Alpha Centauri stars and so it was probably bound to them by gravity.

---

* The same technique, known as blink comparison, is widely used to discover asteroids in the solar system. Its finest hour came in 1930 when Clyde Tombaugh used it to discover Pluto.

By 1917, Innes had attempted a parallax measurement (see 61 Cygni) for the new star, and concluded on somewhat shaky evidence that it was closer even than Alpha Centauri. Taking a punt, he proposed that his discovery should be known as Proxima. Fortunately for Innes' subsequent reputation, US astronomer Harold L. Alden, working at Yale Observatory's Johannesburg outpost, made more accurate measurements that clinched the case in 1928.

Even while its precise distance remained uncertain, however, Proxima was clearly something new. By some distance the faintest star on record, it was (in the new-fangled terminology of the Hertzsprung-Russell diagram – see Alcyone) an extreme example of a red dwarf – relatively small, rather cool and incredibly faint.

A few other examples of stars in this class had already been discovered in our cosmic back yard – orange or crimson red and pumping out perhaps a few percent of the Sun's light or thereabouts. But Proxima took things to extremes: in terms of spectral classification it was a dull red M5 star, shining with just 1/20,000th of the Sun's brilliance.

Barnard's speeding star

Just a year after Proxima's discovery, another famous red dwarf came to light – one that is a bit easier to track down and has the benefit of being visible from pretty much everywhere.

Barnard's Star is famous for being the fastest-moving star in the heavens, and the fourth closest star to the Sun at a distance of 5.95 light years. It's named after E.E.

Barnard, who was the first to spot its high proper motion in 1916[2]. Racing across the sky at a pacey 10.3 arc seconds per year, it covers a distance in Earth's sky equivalent to the width of a full moon in about 180 years (compare that to the 467 years Proxima takes to cover the same distance).

Barnard's Star is also considerably brighter than Proxima. At magnitude 9.5 you should be able to see it with decent binoculars, and because it's some distance outside of the stream of the Milky Way, there are fewer confusing background stars.

The star lies in the large but rather faint constellation of Ophicuhus, the serpent bearer (see page 182). Look towards Ophi's left shoulder and the relatively bright star Cebalrai. This forms the northwest corner of a lopsided

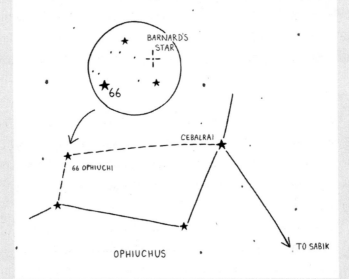

quadrilateral with three fainter naked-eye stars – our next step is to hop to 66 Ophiuchi at the northeastern corner. Through binoculars you should be able to see that 66 is itself one corner of an equilateral triangle with two fainter stars (magnitudes 8.0 and 8.8) to its north and east. The triangle is a little wider than a full Moon, and Barnard's Star currently sits a little way outside its eastern edge, closer to the fainter of the two corner stars.

At first, it seemed hard to believe that such faint stars could possibly exist, but many more soon came to light as telescopes and search techniques improved. This surge in red dwarf discoveries came at an opportune moment in the wider story of astronomy, so forgive a short but important diversion into the crucial question of the brightness and mass of stars...

Based on everything we've seen so far, astronomers knew by the early twentieth century that there was a vast variation in the intrinsic brightnesses of stars and that this was related, in most cases, to their surface temperature and colour through their position on the Hertzsprung-Russell diagram. Because a star's surface temperature depends on how much escaping radiation is heating how big an area, they could also work out a ballpark figure for the diameter of stars with a known luminosity. In Proxima's case, for instance, this works out about one-sixth of the Sun's diameter, or about 50% bigger than the planet Jupiter.

But how much material did a star of a certain size contain? Was it packed in tightly or spread out in a tenuous cloud? Again, you could make certain assumptions: just as Arthur Eddington had cracked the link between a star's energy output and its size by considering the balance of outward radiation pressure versus inward gravity, you could reasonably assume that brighter stars with greater outward pressure would be less dense and that fainter stars would "pack more in".

Putting some observational flesh on these rather bare theoretical bones would be crucial to understanding what the H-R diagram really meant, and indeed to understanding the structure of stars from the biggest bluest giant to the humblest reddest dwarf – so it's fitting that the crucial work once again came down to Ejnar Hertzsprung. In 1923, he published an ingenious study that compared the estimated masses of stars in binary systems[*] with their intrinsic luminosity calculated from parallax[3]. A clear, though imprecise, link between mass and luminosity emerged – the more massive a main sequence star is, the more brightly it shines, and the luminosity increases much more sharply than mass[†].

Ultimately, this comes down to the nuclear physics we explored briefly on our trip inside the Sun: the temperature and pressure in a star's core depend on the amount of material pressing down from above, and the rate of fusion reactions is *very* sensitive to variations. We'll see when we visit Sirius that hotter, denser and more massive stars can make use of a completely different reaction chain to pump out energy at an accelerated rate, but the crucial takeaway for our present purposes is that a red dwarf like Proxima has about an eighth of the Sun's mass but shines

---

[*] As you might recall from Mizar, binaries offer you a variety of ways of figuring out the likely mass of the components, depending on how much information you have.

[†] In 1924, Eddington worked out the theoretical relationships involved, and as a result he, rather than Hertzsprung, is usually credited with discovering this "mass-luminosity relationship".

with just 0.00005% of its brightness. Once you start trimming down the size of stars, their luminosity rapidly falls off a cliff. A shift towards longer, redder wavelengths as the star's temperature drops helps to exaggerate the fall in brightness. Proxima's *overall* energy output is actually a slightly more respectable 1/500th of the Sun's, but much of the energy that it does emit is shed at invisible, infrared wavelengths.

There is a lower limit, however – once you get down to about one-twelfth of the Sun's mass, the heating and compression forces are just too weak for fusion to ever ignite properly. This is the cut-off point for stars, and objects below this threshold are classed as a different type of object, called a brown dwarf. Although their existence was first predicted in the 1960s, it wasn't until 1995 that astronomers finally tracked down one of these "failed stars" – an object called Teide 1 in the Pleiades star cluster. The increasing scope and sensitivity of infrared sky surveys has since led to the discovery of hundreds more, especially within star-forming nebulae. The same surveys have also confirmed what was hitherto only suspected – red dwarfs throng the infrared skies, perhaps accounting for three-quarters of all stars in the Milky Way, while remaining almost entirely out of sight unless they happen to lie on our cosmic doorstep.

★ ★ ★

But while red dwarfs might be small and faint, it's still wise not to underestimate them. Stars such as Proxima can pack a surprising punch – a fact first unearthed by Dutch-American Willem Jacob Luyten, a former student of Hertzsprung's who made a career out of discovering and studying the sky's faintest stars at the University of Minnesota. In 1948, Luyten discovered a pair of evenly matched red dwarfs, some 8.7 light years away in the constellation of Cetus, and while checking in on them a few months later, he observed a sudden increase in brightness from the fainter of the pair, lasting for just a few minutes. Further outbursts were soon spotted in this star (later given the variable identifier UV Ceti) and others. Astronomers realised they were looking at an entirely new type of object, a "flare star."[4]

In 1951, Proxima itself joined the ranks of the flare stars thanks to a close study of 590 photographic plates at Harvard. These showed dozens of brief outbursts, sometimes increasing Proxima's brightness by up to two and a half times or even more.[5]

So red dwarfs might be feeble in the grand scheme of things, but they're prone to huge outbursts that come and go in a matter of minutes. What sort of mechanism can cause something like that? Well, a big clue lies in the name "flare stars" – astronomers think that the process at work is similar to the one that causes solar flares on the Sun. American astronomers Milton Humason and Alfred Joy, gathering a spectrum from UV Ceti at California's Mount Wilson Observatory a few months after Luyten's discovery, caught one of the outbursts in action and discovered that not only did the star's brightness increase (in this case by a factor of 40), but its entire light output shifted to hotter, bluer wavelengths. While UV Ceti's normal surface temperature was around 3,000°C, during an eruption it soared to 10,000°C or more.[6]

How can feeble dwarfs give off so much more energy from their flares bigger stars like the Sun? The answer seems to lie in an internal structure that allows them to generate more powerful magnetic fields. While the Sun has three internal layers (core, radiative zone and convective zone), a red dwarf has just a core and a deep convective zone that carries heat all the way to the surface*. What's more, with less radiation pressure to support them against gravity, dwarfs are much denser than other stars: Proxima has an eighth of the Sun's mass, but it is only half as big as Jupiter, and therefore 33 times denser than the Sun on average. This turns dwarf stars into seething cauldrons of hot gas.

Any star's magnetic field is generated by churning currents of electrically charged material in its convective layer, so perhaps we shouldn't be so surprised that the deep, dense convection cells of Proxima and similar stars have such dramatic effects. Compared to the Sun, many red dwarfs show sunspots and flare activity that are turned up to 11. Surfaces ruptured by emerging loops of tangled, intense magnetic fields, are covered by vast, changeable starspots – blotches so big that they can affect the star's overall brightness as they rotate and the spots are revealed or hidden from Earth.

And just as we saw on the Sun, these fields can sometimes short-circuit and reconnect, with the intense magnetism involved unleashing vast amounts of energy that put larger stars in the shade. In 2008, for instance, NASA's Swift satellite – designed to monitor high-energy gamma rays from violent events in distant galaxies – instead detected an outburst from a nearby young red dwarf star called EV Lacertae, thousands of times more powerful than even the biggest solar flares such as the Carrington Event.

Proxima's flares aren't quite in that league, however. It seems

---

* This also makes red dwarfs the only stars with properly "mixed" interiors: currents inside the star can carry away helium generated by nuclear fusion in the core, and replace it with fresh hydrogen fuel from the upper levels (in more massive stars the radiative zone squats on top of the core and prevents this sort of churn, putting a severe limit on the core's potential fuel supply). Along with a naturally sluggish rate of fusion, this renders red dwarfs functionally immortal, with theoretical lifespans of trillions of years.

that the most violent flare stars are also the youngest and fastest spinning: the star's magnetic power plant works a bit like a bicycle dynamo – the faster you pedal, the more impressive the result. However, all this surface activity also results in the star losing mass through a strong stellar wind, draining angular momentum and slowing the spin right down over a couple of billion years. Therefore, while EV Lac turns every four days, the more mature Proxima (a shade older than the Sun at 4.85 billion years old) spins on its axis just once every 82.6 days and this produces relatively sedate flares.

The precise strength of Proxima's activity has a vital bearing on the habitability of the planet we now know to be orbiting the star. We'll learn a lot more about so-called extrasolar planets and the methods of finding them in the next chapter, but for now it's enough to say that this alien world, confirmed in 2016 by an international team, has a mass of about 1.3 Earths and orbits its star in just 11.2 days.[7]

That orbit is shorter than anything in our solar system, but it puts the new planet, generally known as "Proxima b", squarely in the Goldilocks zone for its red-dwarf parent star – the region that's "not too hot, not too cold, but just right" to sustain liquid water on its surface. And where there's liquid water, perhaps there could be life?* That's much less likely if the planet's surface is being periodically bombarded with fast-moving solar wind particles, deadly X-rays and high-energy ultraviolet radiation – hence the level of interest in Proxima's activity. Early hopes

---

* Some people think that treating liquid water as a prerequisite for alien life is an example of Earth chauvinism, but there are some good reasons why it's not. If complex chemicals are ever going to get together and kick off biochemistry, they need a medium to move around in, and water is the most supportive, stable and plentiful solvent we know of.

hinged on Proxima's advanced age, relatively modest activity and an unknown factor – the hope that a rocky planet bigger than Earth would generate a decent magnetic field of its own, capable of protecting the planet from the worst effects of life so close to its star.

However, the prospects for life on Proxima b suffered a blow in 2018 when astronomers studying data from the University of North Carolina's Evryscope – a revolutionary high-resolution camera designed to continuously monitor large areas of the sky – reported a "superflare" from Proxima. During this hour-long eruption, the star's visual brightness increased by a factor of almost 70, bringing it to the threshold of naked-eye visibility[8]. It's quite possible that such flares occur several times a year and have previously been missed for want of looking. If so, then while that's potentially fatal for the chances of life in the next-door solar system, it does at least slightly boost your hopes of tracking down our elusive stellar neighbour.

10 – HELVETIOS

In search of strange new worlds

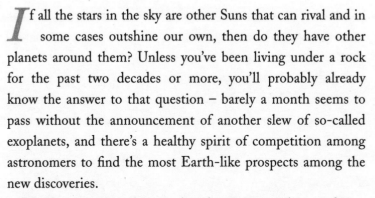

*I*f all the stars in the sky are other Suns that can rival and in some cases outshine our own, then do they have other planets around them? Unless you've been living under a rock for the past two decades or more, you'll probably already know the answer to that question – barely a month seems to pass without the announcement of another slew of so-called exoplanets, and there's a healthy spirit of competition among astronomers to find the most Earth-like prospects among the new discoveries.

But the surprising thing is that the entire exoplanet industry only kicked off in earnest as recently as 1995, after generations of false starts and erroneous claims. And the star that started it all is next on our list.

To the naked eye, Helvetios is a faint dot of light in the constellation of Pegasus, the winged horse. Largely overlooked and nameless for centuries, it was picked up by British Astronomer Royal John Flamsteed during his decades-long trawl of the sky around the turn of the seventeenth century, and was soon after blessed with the "Flamsteed number" of 51 Pegasi*.

Pegasus is a large and obvious constellation, visible in evening skies across the world between September and January, and in morning skies from around May/June onwards. However if you're looking for a strong resemblance to a horse, you'll have to squint very hard – the constellation's brightest stars form a large empty square that is supposed to be the body, with chains of fainter stars emerging from the two western corners to mark out the neck, head and forelegs (of the hind legs and wings, there's no sign). In terms of this star-picture, Pegasus is actually the "wrong way up" for northern-hemisphere observers, with the head to the southwest and the feet sticking up to the northwest. South of the equator, on the other hand, the horse appears the right way up, making a majestic leap across the northern sky.

Whichever part of the world you're looking from, Helvetios defies its catalogued magnitude of 5.5 to be a pretty easy spot under dark skies. Simply identify the western pair of stars in the famous Square of Pegasus (the red giant Scheat and the blue-white Markab), and track down the line between them. Helvetios is halfway along the gap and slightly west (outside) of the square itself.

Taking a look at Helvetios through binoculars or a telescope doesn't reveal anything much out of the ordinary – a single yellowish star that is a pretty good match for our own Sun. Lying 50.45 light years from Earth, it's actually about 36% brighter than

---

* Despite the name, the idea of applying numbers to the stars in Flamsteed's catalogue was actually implemented not by J.F. himself, but by comet-botherer Edmond Halley in an infamous "pirate" edition of 1712. The numbers went through several iterations before settling on the form we know.

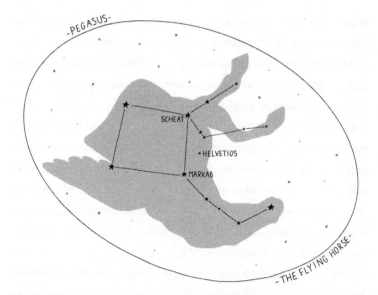

our own star, and because its surface is actually a couple of hundred degrees cooler than the Sun's, we can estimate its diameter at 1¼ times that of the Sun or about 1.7 million km across. All of this suggests Helvetios is a pretty average main-sequence star, with about 4% more mass than the Sun. Clues in the stellar spectrum, meanwhile, suggest it's also a little more advanced in age at about 6 billion years old.

So about that planet…

★ ★ ★

The idea that other stars might have planets around them goes back at least to Giordano Bruno, an Italian monk and philosopher who, in 1600, got himself burned at the stake for a colourful range of heresies including a belief in reincarnation, doubts over the divinity of Christ, and arguing for the "plurality of worlds" – the idea that there are other habitable planets besides Earth.

When the invention of the telescope less than a decade later led eventually to a wider acceptance that stars were just other Suns a long way off, the idea of other solar systems also became a common assumption for astronomers and philosophers – or at least among those willing to set aside religious concerns.

However, the possibility of observing these worlds was another matter entirely. Since planets shine only by reflected starlight and are tiny compared with stars, you don't even need the back of an envelope to realise that they will be immensely fainter than any star. Even if you did have a telescope big enough to gather the required light, you'd find yourself fighting to see anything in the glare from the star itself.

If direct observation was out of the question, then attempts to find exoplanets would have to rely instead on indirect methods, detecting the presence of planets from their influence on the stars they orbit. This is more difficult than it sounds – while a planet might in theory wield a gravitational influence on its star, the effect of that influence depends crucially on the difference in mass between the two objects, and, more significantly, on the distance between them. A planet and its star exert an equal and opposite force on each other but the same force can have a much bigger effect on an object with smaller mass than on one with larger mass. This means the planet moves a lot as it orbits under the influence of gravitational attraction, while the star stays almost still.

Both star and planet are actually orbiting around their shared barycentre or "centre of mass" – the pivot-point of a gravitational seesaw that lies somewhere between their individual centres. The barycentre in any celestial double-act always lies closer to the more massive object, and in the case of stars with planets, where the difference in mass is huge, it almost always sits somewhere inside the parent star. As a result, the star slowly "wobbles".

It was on this wobble that astronomers first pinned their hopes of finding planets around other stars. The most obvious way to spot a wobbling star is to precisely track its proper motion, or drift across the sky, and see whether it shows signs of being pulled around. This tends to be a lot easier if the star is nearby and moving fast through the sky, so the earliest planet-hunting attempts understandably focussed on the stars with high proper motion that were likely to be our closest neighbours in space. In the early 1840s, German astronomer Friedrich Bessell used this principle to find evidence of unseen companions affecting two of the brightest stars in the sky, but his objects had Sun-like masses and eventually turned out to be hidden stars (we'll pick up this story when we come to Sirius).

Probably the most famous of several early planet-hunting attempts involved Barnard's Star, a nearby, fast-moving red dwarf star discovered in 1916. From 1937, Peter van de Kamp, professor of astronomy at Pennsylvania's Swarthmore College, began a project to take regular snapshots of the star and its surroundings, building up an unparalleled record of its speedy proper motion across the sky. In 1963, based on his studies of more than two thousand photographic plates, he announced that Barnard's Star wove a slightly drunken path across the sky as it was tugged this way and that by the influence of a planet 60% heavier than Jupiter, in a 24-year orbit[1].

At first glance, Van de Kamp's discovery seemed to stand up scientific scrutiny – the wobbles he reported certainly appeared to be real, although he changed his mind on several occasions about what exactly they represented. The Dutchman's 30-year dataset was hard for anyone to replicate independently from a standing start, and so it was not until the mid-1970s that others began to report how, despite their own lengthy observations, they simply weren't seeing the same wobbles. Today, the entire affair is usually blamed on systematic errors in the telescope mounting at Swarthmore's Sproul Observatory (where Van de Kamp served as director from 1937 to 1972), although he stood by his claims until his death in 1995.

The Barnard's Star saga left other astronomers with a bitter taste in their mouths, and from the 1980s a new sceptical attitude took hold – not just about claims of detection from stellar motions, but about the entire possibility of extrasolar planets. Perhaps our solar system was a rarity, and other planets simply weren't there to be found?

And then, in 1983, the Infared Astronomical Satellite (IRAS) was launched. As the first permanent infrared telescope in space, one of its big successes was the discovery of a number of young stars that appeared brighter than expected in the infrared. This so-called "IR excess" suggested the presence of large amounts of warm dust around them. Images of a young nearby star called Beta Pictoris, meanwhile, confirmed the existence of a large flattened disc in orbit around it. Both of these signs suggested that other stars naturally gave birth to planetary systems, although actual planets remained conspicuous by their absence.[*2] An obscure star in Pegasus would soon change all that.

Russian-American astronomer Otto Struve had put his

---

[*] One curious exception was the zombie solar system discovered in 1992, thanks to tiny changes in the timing of a metronomic celestial clock called a pulsar (see the Crab Pulsar for more on these strange objects). This tiny stellar remnant is being pulled around by three planet-sized objects, probably created from debris left behind in the aftermath of a supernova explosion.

finger on a solution to the exoplanet challenge as early as 1952, but as is often the case, astronomers had to wait for technology to catch up. Struve, who had specialized in the study of multiple star systems, pointed out that as well as creating a side-to-side wobble, orbiting planets would also tug their star towards and away from Earth, creating small Doppler shifts in the wavelength of starlight similar to those seen in spectroscopic binaries such as Mizar. By measuring these red and blue shifts of starlight, and filtering out other effects, he theorized that it should be possible to detect the influence of planets on the star's "radial velocity" – its straight-line speed towards or away from Earth.

A big advantage to this radial velocity method was that it could work on light from any star bright enough to form a decent spectrum, regardless of its distance from Earth. The major problem, however, was that the likely variations in velocity from even the most massive orbiting planets would be measured in metres per second; the resulting Doppler shifts would be *tiny* compared to those commonly caused by a star's normal motion through space.

Measuring such tiny shifts required spreading out the spectrum of starlight more widely than had ever been done before, and it was not until the early 1990s that the first instruments could be built that were up to the job. The basic principle of a so-called Echelle spectrograph is to split starlight apart once into a broad rainbow spectrum, and then split each colour of that rainbow apart again using a second optical device (a precisely formed diffraction grating). This had been done for sunlight and other bright sources since the 1940s, but it was only in 1994 that ELODIE, an Echelle spectrograph suitable for splitting faint starlight and detecting exoplanets, was installed at the Observatoire de Haute-Provence in France. The key to the instrument's sensitivity was to use an optical fibre that delivered light directly from the focus of the observatory's biggest telescope. This not only cut out several stages in the light's journey, but also allowed the spectrograph to be isolated in a stable environment instead of swinging this way and that on the end of the telescope mount.

ELODIE was the brainchild of Swiss astronomer Michel Mayor and his graduate student at the University of Geneva, Didier Queloz. Mayor had co-designed a previous instrument, CORAVEL, which had identified low-mass companions around several stars in the solar neighbourhood, and the new spectrograph was ostensibly intended to answer the question of whether these objects were red or brown dwarfs.

Planet-hunting was more or less an added bonus, but as Mayor and Queloz worked through a list of interesting targets, they spotted something unusual in the spectrum of 51 Pegasi – a periodic wobble with a maximum value of 55 metres per second in either direction that repeated every four and a quarter days. Accumulating the necessary observations to show that the wobble was a statistically valid measurement took several months, and still Mayor, wisely recalling the fate of previous claims, elected to sit tight on the data until the following season. Thus, when Pegasus wheeled into view again in mid-1995, and the oscillations of 51 Pegasi remained the same, they were sure they had their exoplanet.[3]

The announcement of 51 Pegasi b, as it was initially known, was greeted with great excitement. In the years that followed, ELODIE and similar instruments around the world began to discover more exoplanets, in a trickle that soon swelled to a steady stream. By the turn of the millennium, some 27 were known.

It was not until 2002, however, that the floodgates began to creak open as the first exoplanet was discovered using a different tool known as the transit method. Simple in principle, but fiendishly tricky in practice, this relies on spotting the giveaway dip in a star's brightness that happens when a planet, whose orbit happens to pass in front of it, blocks out some of its light. As astronomers have found ways of putting transits to work in increasingly ambitious satellites such as France's COROT (2006–12), and NASA's Kepler (2009–18), they've become a major source of exoplanet discoveries.

★ ★ ★

The discovery of a planet around Helvetios catapulted it from obscurity. The star received its proper name (in honour of Mayor and Queloz' Swiss homeland) when the International Astronomical Union dished out its first tranche of exoplanet nomenclature in 2015. The planet itself, meanwhile, is officially known as Dimidium – a rather dull reference to the fact it has about half the mass of Jupiter[*].

The most remarkable aspect of Dimidium was also the first thing that its discoverers spotted, and something that initially rang alarm bells for sceptics – its remarkably short orbital period. Orbiting in just 4.23 days, the planet skims just 7 million km above the searing-hot photosphere of Helvetios (for comparison, Mercury stays 45 million km from the Sun even at its closest).

---

[*] For a while it looked like the rather more picturesque name of Bellerophon (the Greek hero who rode Pegasus) was going to catch on, but the IAU hath spoken.

Dimidium's close orbit and high mass* amplify its gravitational tug on Helvetios, making it easier to detect. The short orbital period that goes along with this made it easier to detect *quickly* (planets with longer periods and/or smaller wobbles require a lot more patience to verify). However, they raise the obvious question of what the heck a large, gas giant planet is doing so close to its star.

There are very good reasons why giant planets should initially form only beyond the safety of a solar system's "frost line" (the region where the easily-melted materials that make up the bulk of a planet-forming disc can remain stable instead of evaporating and being blown away into interstellar space). It's true that these are mostly based on modeling of radiation, solar wind and gravitational effects in our own infant solar system, but there's no reason why they shouldn't apply equally around other stars.

Yet, Dimidium soon turned out to be the standard-bearer for a whole new class of planets in this sort of orbit – now known as "Hot Jupiters". Some of these giants, which may reach masses up to about ten times that of Jupiter itself, have atmospheres that are actively boiling away into space, leaving a trail of gases that cross the face of their star and imprint their own absorption lines onto its spectrum. In a grim foreboding of Dimidium's fate, astronomers have even found burnt-out Earth-mass husks called Chthonian planets†, which appear to be the exposed cores of

---

* Firmly established as 0.46 Jupiters by other means in 2016: the radial velocity method only catches the element of the planet's gravitational influence that moves the star towards or away from Earth, not up and down or from side to side. So, depending on the tilt of its orbit relative to Earth, a planet's mass can vary significantly – radial velocity on its own just gives you a *minimum*.

† From *khthon*, an ancient Greek term for the Underworld, domain of Hades.

what were once gas giants. If these planets planets did initially form in the regions beyond their solar system's frost line, then clearly something must have caused them to spiral in towards their suns later in their development. Currently, the smart money is on some form of braking caused by interaction with gas left behind in the planet-forming disc.

If nothing else, the variety of strange new worlds being discovered around other stars should maybe teach us to count our blessings. As they rode the spiral track towards their current orbits, the migrating hot Jupiters most likely disrupted the orbits of any rocky, potentially habitable worlds closer to their stars. With increasing evidence from computer models of our own solar system that Jupiter went through its own period of happy wandering early in its history, we should be grateful that it did not end up following Dimidium's example.

11 – ALGOL

Variable stars and hidden binaries

＊

We've touched on the subject of stars that change brightness at a couple of points already – for instance when we saw the spectacular flares emitted by Proxima Centauri, and the unpredictable variations associated with newborn stars such as T Tauri. But when you look closely, an awful lot of stars in the sky turn out to be variable in one way or another.

We're not talking here about the familiar nursery rhyme twinkling that you'll see from most stars from time to time, and which at its most extreme can turn a bright star into a multi-coloured dancing disco ball as you try to get a focus on it through binoculars or a telescope. That sort of twinkle is entirely down to Earth's turbulent atmosphere, which can act like a combined magnifying lens, prism and general pain in the butt, bending starlight this way and that and splitting it into different colours, to the frustration of stargazers everywhere*.

---

* Pro tip: Twinkling is more of a hindrance when trying to see objects near the horizon, since you're looking through a thicker layer of atmosphere. And it only affects stars because their light is concentrated into a point; the naked-eye planets, which are effectively tiny discs (even if you can't see them as such), are pretty much immune.

This is why the professionals site their telescopes on top of the nearest convenient mountain.

No, instead we're interested in actual variations in the brightness of stars, which come in a variety of flavours. Some may go through regular pulsations on cycles that vary from minutes to years; others experience occasional outbursts (predictable or otherwise), and still others display abrupt dips in brightness followed by equally sudden recoveries.

Algol was *probably* the first of these stars to be discovered – though it depends who you ask. The earliest variable star to be described as such in the west is Mira, which we'll be visiting shortly, but Algol's name and its prominent position in the star pattern of Perseus (and its more ancient precursors) all suggest that its curious behaviour was spotted in ancient times, even if a definitive statement remains elusive.

The constellation Perseus sits in the midst of the northern Milky Way. For skywatchers at mid-Northern latitudes, it skims the northern horizon on evenings around May and June, but passes high overhead in November and December. Around this time, it's best seen over the northern horizon for southern-hemisphere stargazers.

Although the constellation is rather shapeless (with perhaps a hint of athleticism in its springing chains of stars), it's pretty easy to find – just look due north of the Pleiades, or southwest of Cassiopeia, the obvious W-shape that acts as counterweight to the Plough on the opposite side of the northern pole star Polaris. Algol is the constellation's second brightest star, lying about halfway along a direct line from the Pleiades to the first star of Cassiopeia's W.

Perseus, perhaps unsurprisingly, represents the Greek hero of

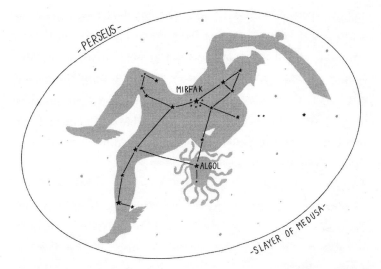

the same name – the one whose legend was lovingly ripped off
for the 1981 Ray Harryhausen classic *Clash of the Titans* (and
its charmless 2010 CGI remake). The Perseus legend is worth a
quick recap since it ropes in so many other constellations in this
area of the sky. So, once upon a time...

Perseus was born under one of those all-too-common Greek
prophecies of doom – in this case, that he would one day kill
his grandfather, King Acrisius of Argos*. Acrisius locked up his
daughter Danaë to keep her away from potential impregnators,
only for the habitually randy king of the gods Zeus to, ahem,
appear to her as a shower of gold. The king balked at the prospect
of cold-blooded infanticide and instead cast both mother and baby
out to meet certain death at sea – where they were predictably
rescued by Dictys, a passing prince from a nearby island.

---

* The early Greek city state, not the stubby-pencil catalogue retailer (though would
you believe that Green Shield Stamp magnate Richard Tompkins was on holiday here
in the 1970s when he came up with the idea for Argos-the-shop?)

Time passed, and Dictys raised the boy and wooed his mother. But Dictys's brother King Polydectes also had designs on Danaë, and Perseus got in the way of the royal prerogative once too often. Tricked at a party into promising King P any gift he cared to name, Perseus was tasked with delivering the head of Medusa (the only mortal among the three snake-haired Gorgon sisters, whose gaze could turn you to stone).

Shenanigans ensued (another set of three monstrous sisters, gifts from the gods, etc.), before Perseus eventually lopped off Medusa's head using the famous "reflection in a polished shield" trick. But as he flew back home on his winged sandals*, his attention was diverted by an attractive young lady tied naked to a rock by the sea.

Andromeda, daughter of King Cepheus and Queen Cassiopeia of Aethiopia, had been left out as bait for the terrible sea monster Cetus on the advice of a helpful oracle. The monster had been ravaging the land and slaughtering all-comers, but Perseus defeated it with a strategic flash of Medusa's petrifying noggin. Andromeda (who is, to be frank, rather lacking in agency compared to some other classical heroines) fell conveniently into his arms, leading to a hasty wedding before Our Hero flew home to give Polydectes his own well-deserved dose of the Medusa touch. The final act varies depending on whose account you follow, but often puts King Acrisius on the receiving end of a stray discus accidentally misthrown by his grandson in a sporting contest.

The Perseus myth unites at least five major constellations – six if you include Pegasus, said to have sprung fully formed from the

---

* On loan from Hermes, the fastest – and clearly the most fabulous – of the gods.

stump of the Gorgon's neck. And at the heart of the story lies Algol, for it represents the eye of Medusa herself.

Algol's association with trouble spread far and wide: for ancient Hebrew astronomers, it was seen as Satan's Head, or associated with the distinctly witchy figure of Lilith (sometimes said to be the first wife of the Biblical Adam). Astrologers viewed it as the most turbulent star in the sky, and one group of researchers has even claim that its periodic variations match up with an Egyptian calendar of lucky and unlucky days from around 1200 BCE. The modern name comes from the Arabic *ra's al-ghul*, meaning head of the demon or ogre.*

Given all this circumstantial evidence, it's more than a little frustrating that no ancient writer we know of specifically wrote about Algol's fluctuating brightness, and some notable authorities seem to have missed it. Most frustrating among these is probably al-Sufi, a tenth-century Persian who didn't so much translate Ptolemy's dusty second-century *Almagest* star catalogue, as strip it down to the chassis and rebuild it with new bodywork and shiny alloy rims based on his own more accurate observations.

However, it seems that astronomers before the telescopic age simply had a blind spot for changing stars – perhaps because acknowledging them would call into question the assumption (which pretty much everyone had inherited from Aristotle) that the heavens beyond the Moon were perfect and unchanging†. The

---

* A name that may ring bells with fans of Batman.

† Comets were written off atmospheric phenomena, while supernovae – brilliant "new stars" that would really have set the cat among the dogmatic pigeons – were rare enough not to be an issue. In fact, the astronomical revolution attributed to Copernicus (d. 1543) only really gained momentum after many people witnessed the Milky Way supernova of 1572 and the Great Comet of 1577.

first person to put Algol's wobbly nature on record, therefore, was Italian lawyer-turned skywatcher Geminiano Montanari *.

From the 1660s, Montanari – a disciple of Galilean science at a time when it was still a touchy subject in Italy – scoured the sky for stars that were not living up to Aristotle's rigorous ideals. Around 1666, he noted for the first time that Algol's brightness was occasionally prone to significant dips: in modern terms it usually shines at a steady magnitude 2.1, but can dip to 3.4.

Montanari continued to monitor Algol for more than a decade until lucrative but toxic appointments at the Venetian arsenal and mint, notorious for their lax attitudes to workplace health and safety with dangerous chemicals around, robbed him of his eyesight. By this time, he had also identified about a hundred other variable stars, yet it seems he never came close to spotting the periodic nature of Algol's winks. Neither, apparently, did his fellow Italian Giacomo Maraldi, an émigré to Paris who paid particular attention to Algol a couple of decades later. It was therefore almost another century before the pattern in Algol's behaviour was finally spotted and its cause identified – a discovery usually attributed to the remarkable but unfortunate John Goodricke.

A scion of low-ranking Yorkshire nobility, Goodricke was struck deaf by illness in early childhood and later educated at the pioneering Academy for the Deaf and Dumb, recently established by Thomas Braidwood in Edinburgh. Returning as a bright 16-year-old to the family home in 1780, he soon fell into company with the new neighbours – noted astronomer Nathaniel

---

* Montanari deserves to be more famous – if only for his 1685 stunt of publishing an "astrological almanac" full of entirely random predictions, in order to show up his more credulous rivals.

Pigott (who had made his name measuring the 1769 transit of Venus in front of the Sun), and his son and oft-uncredited observing partner Edward.

Despite a decade's age gap, Pigott Junior struck up a friendship with Goodricke, and was soon introducing him to the wonders of modern astronomy. The pair began a scientific partnership, with Pigott suggesting a number of variable stars that might be worth Goodricke's time to observe – including Algol. At the time, most variable stars were thought to be like Mira – the star we'll meet in the next chapter, which pulsates in a cycle lasting hundreds of days. Through systematic observations, Goodricke realised by 1783 that Algol had a much shorter cycle – just 2 days, 20 hours and 45 minutes to be precise. Algol's change in brightness, when it came, was a sudden dip rather than a slow decline, with an equally rapid recovery about ten hours later.

As the young researchers looked for for an explanation, it seems to have been Pigott who initially suggested that Algol was being eclipsed by a fainter object that periodically stole some of its light. But in a remarkable act of scientific altruism, the older man encouraged his protégé to develop the idea and officially announce the discovery via a letter to the Royal Society read in May 1783[1].

Within weeks, other astronomers had turned their gaze towards Algol, confirming the metronomic nature of its changes. Goodricke's observations soon saw him awarded the Royal Society's prestigious Copley Medal – its youngest ever recipient at the age of just 19. His death from pneumonia in 1786, on the eve of his election as a society fellow, robbed British astronomy of a precocious talent.

While Goodricke's lasting fame is for the discovery of what we now call an eclipsing binary system, the truth is a little messier. With no other evidence to suggest that Algol was a binary star, Goodricke hedged his bets and also suggested an alternative explanation – the presence of dark starspots that periodically came into view on the star's surface.

As we'll see, this idea had already been proposed as a potential explanation for the rhythmic variations of Mira, and despite not really working for cases like Algol (why would spots be visible for just 10 hours of every three days, rather than half the time as each hemisphere rotated in and out of view?) it became the standard explanation for all variable stars through much of the nineteenth century. Further challenges to the eclipse theory soon arose in the form of variables that made gradual, rather than sudden, transitions from one brightness level to another, and another blow came in 1843 when the astute Prussian astronomer Friedrich Wilhelm Argelander detected a small but undeniable change in the period separating Algol's dips in brightness.

It took until the 1880s for someone to resurrect the idea of Algol as an eclipsing system, when American astronomer Edward Pickering put in the hard yards to calculate various possible orbits and show how the star's apparent brightness throughout its cycle could be predicted to within a fraction of a magnitude. The changing period detected by Argelander, he showed, could be explained if you also allowed for the influence of a third, more distant partner slowly tugging on the central pair to produce a 32-year oscillation.[2] A decade later, Germany's Hermann Carl Vogel, found conclusive evidence, in the form of Doppler shifts affecting the wavelength of light emitted from Algol as the brighter star is pulled in various directions by its orbiting companion.[3]

Light from the secondary star, however, remained frustratingly elusive. While spectroscopic binary stars such as Mizar showed spectral lines that split apart and reunited as the two elements of the system moved in opposite directions, Algol's lines simply moved back and forth, indicating that the vast majority of the system's light was coming from a single star.

Another way of measuring the companion's brightness, in theory at least, was to detect the system's "secondary minimum" – the much smaller dip in its overall output caused when the fainter star passes behind the brighter one. But try as they might, no one could spot it. This was a problem in need of a technological fix – in this case, an ingenious new addition to the astronomer's toolbox called the photoelectric photometer.

The photometer works on the same principles as today's solar cells, using a semiconductor material that generates a small flow of electricity across its surface when it is energized by individual photons of incoming light. By measuring the current produced, you can get an exact measurement of a star's brightness from moment to moment. The idea was pioneered by Joel Stebbins, the young Nebraska-born director of the University of Illinois Observatory, in the early 1900s. From 1910 onwards, he began to make regular measurements of variable stars including Algol, eventually confirming that there was indeed a secondary minimum – a drop in brightness of less than 4 per cent as the brighter star eclipses the fainter one.

Photometry using filters for specific wavelengths and colours of light allows you to pull off some clever tricks. For instance in 1939, John Scoville Hall at Amherst College compared measurements of Algol using an infrared photometer with Stebbins's own results from the blue end of the spectrum.

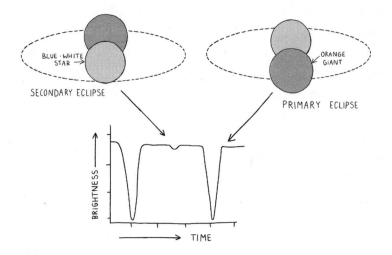

By showing that the primary eclipse is less significant in the infrared and the secondary minimum stronger, he proved that the companion's light is biased towards the infrared.[4] If you remember the relationship between star colour and temperature from our visit to Aldebaran, you'll realize this means the secondary is much cooler than the primary. This was confirmed in 1978, when astronomers from the University of Texas in Austin finally managed a direct measurement of the secondary's faint absorption spectrum.[5]

So what exactly do we know about this system? Nearly all of Algol's light, it seems, comes from a hot blue-white star with more than three times the Sun's mass and 180 times its luminosity, at a distance of about 90 light years from Earth. The cool yellow-orange companion, meanwhile, weighs in at about 0.7 Suns, but still pumps out seven times more energy than our own star. What's a star with seven tenths of the Sun's mass doing shining that brightly? Well, it turns out that the companion is an orange giant: three and a half times bigger than the Sun and 30 per cent larger than the bright primary.*

If you've been paying attention, this may set alarm bells ringing: how has a star with less mass than the Sun made it all the way to becoming an orange giant, while its heavyweight sibling – which according to everything we know about stellar evolution should age much, much more quickly – is still happily shining on the main sequence? It turns out that quite a number of other close binary systems also display this "Algol paradox", where the evolutionary states of components with differing masses seem to be the wrong way around.

In the mid-1950s, two astronomers independently hit upon the solution to this conundrum. Czech-born Zdeněk Kopal, head of the Astronomy Department at the University of rainy Manchester, and John Avery Crawford at the sun-kissed University of California, Berkeley, both spotted that the less massive components in each case shared an unusual feature: their diameter precisely matched the limits of their gravitational grasp.[6]

In any system where two or more massive stars are competing for space, their gravitational spheres of influence will inevitably be distorted into teardrop-shaped zones called Roche lobes.† Material within one star's Roche lobe will remain gravitationally bound to it, but if it strays outside, it can drift away into space or even be swept up by a rival star. The fact that Algol's secondary is just the same size as its Roche lobe shows that this is exactly

---

* Just for the record, the third member of the system is a white star, a little more massive than the Sun, that keeps its distance from the central pair in a 680-day orbit.

† Named after nineteenth-century French astronomer and mathematician Édouard Roche.

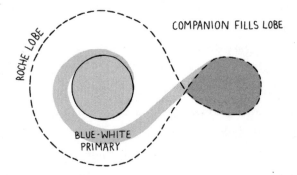

what has happened, and occasional bursts of high-energy X-ray emissions – possibly due to stray gas clouds being heated by the intense magnetic fields around the pair – suggest the process is still going on today.

When the stars of the Algol system were born about 570 million years ago, the see-saw balance of their masses was tipped in the opposite direction to that which we see today. The more massive star burned through its hydrogen fuel supply much more quickly, and eventually began to swell into a red giant – but as it did so, it burst beyond its Roche lobe and lost its grip on its own outer layers. Much of this hot gas spiraled down onto its relatively puny companion, delivering a cosmic protein shake that helped it build body mass and (by increasing pressure onto the core), boosted its own energy output. By the time the more evolved star had been stripped down to its Roche limit and could hold onto the material that remains in its outer layers, the status of Algol's two central stars had been switched.

This curious role reversal, known as mass transfer, turns out to be surprisingly common in cosmic terms, and we'll be seeing similar processes at work when we come to some of the most violent objects in the Universe in a few chapters' time.

There's one final thing that's worth mentioning before we leave this most sinister of stars: by sheer coincidence, Algol may indeed have once wielded a destructive influence over our planet. While today it lies a safe 90 light years from Earth, rewinding the cosmic clockwork by tracing back the motions of nearby stars shows that 7 million years ago, Algol passed within about eight light years of our solar system. Around this time, it would have been the brightest star in the sky, and thanks to the combined mass of its three member stars, it could have disrupted the orbits of icy comets orbiting in the Oort Cloud – a vast halo of comets that surrounds our solar system at a distance of about a light year. One study suggests that more than 1.5 million comets could have been disrupted onto paths that sent them plunging into the inner solar system.[7] The arrival of these infalling comets would have been spread over hundreds of millennia, but collisions with Earth and the other planets would have been inevitable. Any impact craters have long ago been wiped from the face of our resilient planet, but who is to say whether several million years ago, Medusa did more than cast a baleful wink in the direction of Earth?

12 – MIRA

Of red giants and pulsating stars

While Algol and its kin vary in brightness through the intervention of a handy neighbour, many other stars undergo changes that are entirely self-contained. Amongst these, the most famous is the curious star Omicron Ceti, better known as Mira.

Of all the many variable stars in the sky, Mira is (at least some of the time) the easiest to spot. And when it's not, you can at least see the gap where it *should* be. The star marks the neck of the constellation Cetus, the sea monster defeated by Perseus using a quick glimpse of Medusa's head.

For skywatchers north of the equator, Cetus skims across the southern evening skies between October and January (though as with most seasonal constellations, early risers can see it in morning skies a few months ahead of its evening appearance). Southern-hemisphere viewers get to see the monster upside down in their northern skies over a similar period.

Although its name is simply the Latin for whale, Cetus's outline is more like a sea lion than anything, with a large flattened

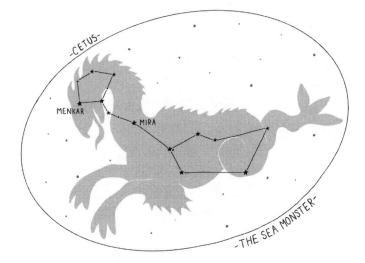

polygon of stars (the body) offset from a smaller, irregular pentagon to its northeast (the head). Mira is usually visible as a red star marking the middle of an elongated neck linking the two. At its peak, it can be the constellation's third-brightest star, easily visible at magnitude 3.04*. However, when the star pulls its periodic disappearing trick, it leaves this particular sea monster as decapitated as Medusa herself. As the star dwindles to minimum brightness around magnitude 10, it can be a tricky spot even for a decent pair of binoculars. Your best bet is to look for a lopsided cross of faint naked-eye stars: one arm runs between 67 and 70 Ceti, the other between 66 and 81 Ceti, and X marks the spot where Mira should be.

While Algol carries a weight of circumstantial evidence to suggest its strange behaviour was known to the ancients, that's not the case with Mira – though you'd think the star's dramatic

---

* Presumably Mira was having an off day when Johann Bayer catalogued it as Omicron, the lowly 15th letter of the Greek alphabet.

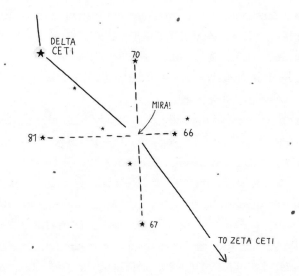

vanishing act at the heart of a fairly prominent constellation would make it even more obvious. However, Mira is its own worst enemy – its 11-month cycle of oscillations means it can remain unseen for months at a time.

Some astro-historians have attempted to link Mira with one or other "guest stars" listed in Chinese chronicles. These unexpected appearances in the sky are usually attributed to novae – distant stellar explosions that we'll explore when we visit RS Ophiuchi – but the steady brightening of Mira might fool someone who only watched it for a single cycle. Unfortunately, the evidence is frustratingly inconclusive.[1]

The discovery of Mira's variability is therefore credited to David Fabricius, a protestant pastor with a parish on the windy coast of Frisia in northern Germany. He first spotted the star in the small hours of a summer night in 1596 while looking for a handy waypoint to pin down his measurements of nearby Jupiter. As he tracked the planet over the course of the next two months, he noticed that his comparison star at first grew brighter, then slowly faded from view. Naturally enough, he assumed that he had found a new nova, but when he spotted the star back in its place in 1609, he realised it must be a different type of object altogether.

In the torrent of new discoveries that followed the invention of the telescope, Mira was subsequently overlooked for more than two decades, before being rediscovered in 1638 by another Frisian stargazer, Johannes Holwarda. Through systematic observations, Holwarda determined that its brightness waxed and waned in a repeating period of about 330 days (since refined to 332 days).

Thereafter, the star languished once again until it was finally propelled to fame in 1662, courtesy of the celebrity astronomer of the age. Johannes Hevelius, from across the Polish border in Danzig (modern-day Gdansk), was an early entry to that small but noble tradition of brewing astronomers, keeping his fellow citizens in beer by day, while spending the proceeds on increasingly ambitious night-time activities that required both a clear head and steady eyesight. Rising to become a city councillor and eventually mayor, he built a private observatory that straddled the roofs of three family houses, equipped with ambitious instruments of his own making. Hevelius courted royal support both in Poland and further afield, and corresponded with Europe's great scientific minds. So, when, in 1662, he published his *Historiola, Mirae Stellae* (*Little history of the Wonderful Star*), people took notice.[2]

Hevelius began to observe Mira regularly in 1659, and kept it up for a quarter of a century. Despite this devotion, he seems to have had a love/hate relationship with the star, doubting Holwarda's suggestion of a regular period and sometimes damning it as an "imposter" to be wary of. In the end, his work was most important for giving the star its name, and introducing it to others who would take things further. In particular, Mira became a *cause célèbre* in the long-running culture war between those who still supported the old Aristotelian view of a fixed and perfect divinely ordained Universe, and the heirs of Galileo, who believed the heavens were subject to change under simple laws of physics.

Prominent among the latter group was Ismaël Boulliau, the grand old man of French astronomy. A Catholic convert with a Calvinist upbringing, Boulliau had learned from both Kepler and Galileo. Painstaking research and systematic observations convinced him that Mira's variations did indeed follow a regular cycle, and in a 1667 book, he made the first attempt to explain its changing brightness. After disproving the possibility that Mira followed an orbit that periodically carried it towards and away from Earth, and dismissing the idea that the intensity of its "inner fires" could follow such a predictable routine, he settled on an insightful alternative, suggesting that Mira's surface had several large dark patches, which periodically came into view as the star rotated.*

This theory became the dominant interpretation of all variable stars for the next two centuries, as surveys of the skies became increasingly systematic and more of these changeable stars came to light†. By the late nineteenth century, more than 250 "Mira-type" variables – reddish-coloured stars showing

---

* While Boulliau was wrong in this case, we now know that some other forms of variable do indeed owe their behavior to such "starspots".

† Even John Goodricke's suggestion that Algol was an eclipsing binary system was largely overlooked in favour of the starspot model.

periodic variations of 100 days or more – had been catalogued, amounting to about three in five of all the variable stars known at the time. The rate of discovery accelerated massively from the 1890s, when Williamina Fleming, former housemaid turned astronomer and the first of the famous "Harvard Computers" (see Aldebaran), discovered clues in the dark absorption spectra that could help identify Mira-like stars, and other new types of variable, without all those tedious years of painstaking observation.

* * *

So, just why do Mira and other "long-period variable" stars act so strangely? As it turns out, the mystery couldn't be neatly solved by either convenient starspots or the handy *stella ex machina* of an eclipsing neighbour. Instead, despite Boulliau's reservations, the clues did in fact lie in Mira's inner fires – the incandescent, high-pressure nuclear furnace of its core. Mira is in fact the prototype for a whole family of variable stars that display their internal instability as a badge of honour.

Thanks to the work of Hertzsprung, Russell and others in the early twentieth century, we know that Mira is a star far more luminous than the Sun, with an enormous size that gives it a cool red surface. With a spectral type of M7 on Annie Jump Cannon's ingenious classification scheme, its surface temperature is around 2,900°C (on average – it varies by about 150 degrees either way with Mira's brightness). This makes it a red giant – a more exaggerated version of the orange giant Aldebaran that we met back in Chapter 3. Our statistical best guess at Mira's distance from us is about 299 light years, with an energy output that ranges between about 8,400 and 9,360 times the energy of the Sun.[3]

Hmm – that sounds odd. Although Mira's pulsation period is

a pretty steady 322 days, its extremes of brightness can vary quite a bit – peaks can range from magnitude 2.0 to 4.9, and troughs from about 8.6 to 10.1. Taking the extremes, that's a change in brightness by a factor of over a thousand – so how can Mira's energy output only be varying by about 11%?

The reason lies in that cool red-giant surface. It's so cool, in fact, that a 300-degree shift in temperature can have a dramatic effect on whether it is emitting radiation at visible wavelengths, or in the infrared. When Mira's visibility dips, its infrared output increases and vice versa.

So, we may not have a star altering its power output a thousandfold, but we still have a pretty big mystery. Mira's varying brightness and surface temperature show that it must be swelling and shrinking in size – as more radiation pours out of the star, its gassy outer layers expand and the surface cools. Then as the luminosity falls, gravity pulls back the outer photosphere and the surface heats up. On average, Mira varies between about 332 and 402 times the diameter of the Sun.

In order to understand why some red giant stars vary dramatically while other remain stable, we need to get to grips with what a red giant actually *is* – and in particular the fact that they come in several flavours. In the early twentieth century, astronomers wondered if they might be stars in the early stages of formation. This wasn't such a bad guess (considering that genuine infant stars like T Tauri share some red-giant properties such as large size and high luminosity), but in fact red giants are at the opposite end of their life cycle. After spending most of their lives shining steadily on the main sequence of stellar evolution, changes to their internal power generations have caused them to brighten and expand to enormous size.

For stars such as Mira, with masses fairly similar to the Sun, these changes begin when the supply of hydrogen fuel in its core is used up. At this stage there can be plenty of hydrogen left in the star's upper layers, but the radiative zone (present in all stars above half a solar mass), forms a barrier that prevents material from entering or leaving. In Mira's case, it took about six billion years for the hydrogen in its core to be largely replaced by helium, the principal waste product of the fusion process.

As the exhausted core's nuclear engine stutters and dies, you might expect a star to fade, but instead something unexpected happens. As we saw inside the Sun, a star's internal layers are held in a delicate balance between the inward pull of gravity and the weight of overlying material on the one hand, and outward pressure due to radiation pushing its way out from the core. This means that as the core radiation falls away, the star's upper layers naturally begin to collapse inwards.

This is a tough break for the layer closest to the core, trapped in a stellar vice between the weight of material from above and the residual outward pressure from the core below. Compressed and heated to temperature hotter than the original core, the shell ignites with its own fusion reactions, running at a much faster rate than those which previously powered the core.

By now, we should be used to the idea that higher luminosity means greater outwards pressure, so it's unsurprising that the layers above this shell of burning hydrogen expand, cool, and shift the star's colour towards orange and red wavelengths. The star swells and becomes a red giant.

Using an H-R diagram to compare the surface temperature and luminosity of stars, it's possible to track the evolution of a star as it moves off the main sequence, crawling up the "subgiant branch" and eventually joins the "red giant branch" proper as it continues through its life cycle. In the 1950s, however, when astronomers got a better insight into how stars are distributed in this part of the diagram[4]*, they noticed giants clustered in certain parts and not others. This is because the initial period of growth into a red giant is just the first phase in the star's later evolution.

While the outer layers are busy expanding into a distended envelope big enough to swallow up any nearby planets, the inner core, robbed of its internal radiation support, is slowly being crushed under its own weight. As it dwindles in size, both temperature and internal pressure rise, until eventually conditions are so extreme[†] that helium nuclei collide with enough force to fuse together.

This second wave of fusion spreads rapidly through the core in an event called the "helium flash". We might expect it to boost the red giant to new size extremes, but once again the physics of the stars wrong-foots us. As the helium flash restores the outward pressure of radiation emerging from the core, the hydrogen-burning shell above is forced to expand outwards, losing density, cooling, and consequently throttling back its own energy production. So, surprisingly, despite now having two power sources, the star's overall energy output drops. This in turn causes

---

* By harvesting data for the members of globular clusters – vast balls of stars in orbit around the Milky Way, all effectively at the same distance from us – Halton Arp, Bill Baum and Allan Sandage were able to accurately plot the positions of thousands of individual stars.

† About 100 million°C and a density of 100 kg per cubic centimetre – which makes the 15 million°C and 150 g/cm3 at the heart of the Sun seem quite reasonable by comparison.

the outer envelope to fall back and heat up, so in comparison to the previous red giant stage the star actually becomes hotter and yellower. This brief phase in a star's later life is marked by a sparsely populated band of the H-R diagram known (somewhat unimaginatively) as the "horizontal branch".

But helium fusion in the centre of an ageing star doesn't last for long. The core rapidly fills up with the waste products you get from knocking helium nuclei together – principally the elements carbon and oxygen (without which none of us would be here). As the remaining helium gets more and more thinly spread, the core's power supply falters again – and this time (at least for stars like the Sun and Mira) there's no way back.

Mira has now embarked on its final encore – giving us a glimpse of something our own Sun will go through about seven billion years from now. Robbed once again of internal support, history has repeated itself, with the inner layers collapsing and heating up. Hydrogen fusion in the shell region has been rekindled with renewed intensity, and it's now been joined by a second power source – a layer of helium fusion following it out through the star and feeding on its waste products. Increased radiation has reinflated the outer layers, causing them to become even cooler than they were before, and pushing the star's energy output into extreme red and infrared wavelengths. In terms of the H-R diagram it's now climbed back up and to the left of the horizontal branch, to hang out among a group of stars called "asymptotic giants"*.

The two fusion layers inside Mira and its asymptotic pals exist in a delicate balancing act – the helium shell relies on the

---

* Not to be confused with asymptomatic – an asymptote is a specific type of curve in mathematics, which happens to describe how these stars are distributed.

hydrogen fusion just above to keep it fed, but when it generates too much energy and radiation of its own, it causes the hydrogen shell to expand and temporarily chokes off its own fuel supply. In the long term, this causes the star to go through a series of "thermal pulses" – major increases in size and luminosity, taking place over the course of a few decades and separated by intervals of anything from 10,000 to 100,000 years during which the star gradually subsides.

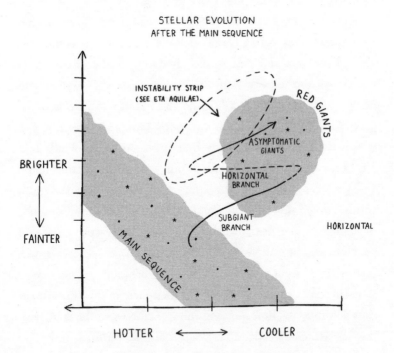

STELLAR EVOLUTION
AFTER THE MAIN SEQUENCE

INSTABILITY STRIP
(SEE ETA AQUILAE)

RED GIANTS

ASYMPTOMATIC GIANTS

BRIGHTER

HORIZONTAL BRANCH

SUBGIANT BRANCH

HORIZONTAL

FAINTER

MAIN SEQUENCE

HOTTER ⟷ COOLER

On a shorter timescale, this instability also gives rise to the short-term pulsations of Mira and its cousins. The exact mechanism is a bit like periodically lifting the lid on a pot of water that is just on the boil, and watching as it subsides. Hydrogen in

Mira's cool upper atmosphere can take on one of two states – transparent and cool, or opaque and warm. When it's warm and opaque, the radiation it traps inside the star caused it to heat up and expand. This cools the outer layers so they turn transparent, and allow heat to escape. The entire star shrinks again, eventually raising the temperature of the upper atmosphere to a point where the lid goes back on, and the cycle repeats. For Mira, this means a steady rise in brightness over a period of about 100 days, followed by a decline that lasts about twice as long.

The instability that red giants go through during these various phases can be enough to break down the quarantine conditions around the core and periodically dredge up some of the waste products from previous fusion – dominated by oxygen and carbon – to the surface. With the star inflated to huge size, its gravity can barely keep hold of its outer layers, and much of this enriched material blows away into space on stellar winds that are hurricane-force compared to our own Sun's gentle breeze. In Mira's case, these winds become visible where they slam into surrounding interstellar gas and heat up to tens of thousands of degrees – the result, visible at ultraviolet wavelengths, is a pronounced shockwave ahead of the star and a comet-like tail, more than 13 light years long, streaming behind.[5]

Evolved red giants shed much of their mass in this way even as they continue to shine, seeding the space between the stars with heavier elements needed to make planets like Earth.

And the Wonderful Star has one final trick up its sleeve – it's not alone. The existence of a companion star was suspected as early as 1918, and the existence of the blue-white Mira B, a little fainter than the red giant at its dimmest, was confirmed in 1923 by prolific Californian binary hunter Robert Grant Aitken[6]. In

1995, when the Hubble Space Telescope turned its gaze on the system, it traced a trail of gas pulled out of the giant star's upper atmosphere towards Mira B. As material spirals down onto the smaller star, it heats up and produces small variations of about 0.2 magnitudes in brightness. These, it seems, are a sign that Mira B is a hot, dense object called a white dwarf – a star that has been through the turbulent red giant phase of evolution already, and moved onto a more settled stellar afterlife. In the next couple of chapters, we'll get to know these curious objects rather better, but <shhh – spoilers!>

13 – SIRIUS
(AND ITS SIBLING)

*The brilliant Dog Star and its
secretive companion*

*And so at last we come to Sirius. Almost twice as bright as any other star in the sky, and outshone only by the Sun, Moon and a couple of nearby planets, the famous Dog Star skims the horizon on northern winter evenings and rides high overhead on southern summer nights, accompanied by its binary companion, the fascinating Sirius B. We've left Sirius until now largely because its elusive companion is the most famous example of a white dwarf – the end state of stars like Mira and our own Sun. But we'll be coming to that in a little while – first let's linger a while at the brighter star (Sirius proper, if you like).

Sirius is situated in the constellation of Canis Major, the larger of the celestial hunter Orion's two dogs (hence its nickname). Lying east of Orion and a little further south, this cosmic hound wisely hangs back behind its master as he picks a fight with the fearsome bull Taurus. The constellation rises and sets a little after Orion, and because it lies south of the celestial equator, northern hemisphere stargazers mostly get to see it through the turbulent lower atmosphere for a few months of the year (it's a fixture of evening skies between December and April). In those same months, our antipodean cousins get the benefit of Sirius high in the sky.

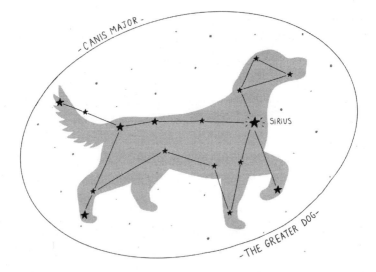

Sirius's name derives from the ancient Greek for "the scorching one". For obvious reasons, plenty of cultures around the world have linked it to particularly powerful deities, but it's more puzzling just how many have linked it to a dog of some sort. Far beyond European influence, Chinese stargazers saw it as the star of the celestial wolf, and the Native American Blackfoot people referred to it as "Dogface". One possible explanation is that the Dog Star got its association because people imagined it following its master – the fairly obvious humanoid figure of Orion.

Another widespread association for Sirius – across the northern hemisphere at least – was as the herald of summer. This might seem odd today, when we're used to greeting its arrival in evening skies from November, but our ancestors, living without artificial lighting, were evidently of the "early to bed, early to rise" school. They welcomed Sirius at its first appearance in the pre-dawn sky an hour or so before sunrise. For most of the Mediterranean and Near East, this so-called "helical rising" acted as an advance warning that the peak of summer would soon be upon them and it was time to break out the sunscreen*. In ancient Egypt some 5,000 years ago, however, it was rather more important. The Dog Star's reappearance in early July was a reliable indicator for the onset of the Nile's flood season – a torrent of muddy waters from the African interior that the Kingdom of the Pharaohs relied upon to sustain its agriculture and prosperity†.

While Sirius's brightness made it a particular fave for astrologers throughout late classical and medieval times, once the Enlightenment came along it was mainly of interest as a potential target for distance-revealing parallax measurements (see 61 Cygni). In 1717, when the famous Edmond Halley rooted through Ptolemy's second-century star catalogue the *Almagest*, he had found that the recorded position for Sirius had drifted by 30 minutes of arc, the width of a Full Moon. Because this large "proper motion", along with Sirius's prominence in the sky, suggested it was probably a nearby star (and thus likely to show a tell-tale shift in direction as Earth orbited the Sun) several failed attempts to measure its position were made over the next century.

---

* Hence the term "dog days", and perhaps also the origin of Sirius's name – phew, what a scorcher!

† The Egyptians unsurprisingly took their worship of Sirius quite seriously – it got its own fertility goddess, Sopdet, and numerous temples were built with alignments towards the point where it sprung back into view.

Sirius remained high on the list of potential nearby stars through to the 1830s, when technology finally caught up with ambition and Bessel and Struve raced to make the first successful parallax measurement (as recounted on our visit to 61 Cygni). However, both ultimately fixed their attentions elsewhere, and so it was left to Scottish astronomer Thomas Henderson to calculate the distance to the Dog Star in 1839[1]. With a distance of 8.6 light years by modern measures, it remained the second-closest known star system (after Alpha Centauri) until the discovery of the first faint red dwarf stars in the early 1900s.

The Dog Star's brightness and distance together imply an overall energy output about 25 times greater than the Sun, while its mass is only about twice as much. This raises the interesting question of why the luminosity of stars can increase quite dramatically for relatively small differences in mass – if you work through the figures you soon find it can't be due to a simple increase in the rate of "standard" proton-proton fusion (the type found in the Sun). So what's going on?

It turns out that if you push a star's mass much past that of the Sun, and the star's gases are seasoned with a few extra ingredients beyond the usual hydrogen and helium, then you can open up a whole new form of hydrogen fusion that runs much faster than the plodding proton-proton process of combining four hydrogen nuclei one step at a time to build helium. Known as the CNO (carbon-nitrogen-oxygen) cycle, it works a bit like a catalyst in a chemical experiment: nuclei of heavier atoms (principally carbon) sweep up individual protons or hydrogen nuclei, and transform first into nitrogen, and then into oxygen. The oxygen nucleus then effectively sheds two protons and two neutrons (a helium nucleus), reverting back into carbon.

CNO CYCLE

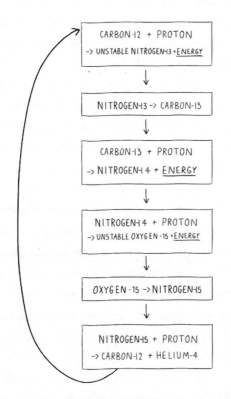

By processing hydrogen nuclei into helium much more quickly, the CNO cycle allows stars like Sirius to generate far more energy each second and shine more brightly. The cycle is also extremely temperature-sensitive, with the reaction rate increasing exponentially once the temperature of the star's core is past a certain level. We'll see in a later chapter how stars like Eta Carinae take this to extremes, but for now it's worth noting that this brilliance comes at a price: despite having more fuel available to burn, stars that use the CNO cycle rip through it at a tremendous pace. They therefore spend much less time shining steadily on the "main sequence" before their core burns out and they begin to show signs of ageing. You know that old saying about flames that burn twice as brightly lasting half as long? Well stars have taken it to heart in a big way.

We'll come back to that ageing process shortly, but first let's return to the mid-nineteenth century. By 1844, Friedrich Bessel was back on the scene with a surprising discovery about Sirius – it was not alone.

Through careful tracking and comparison with observations going back to 1755, Bessel had found that the proper motions of both Sirius and another fast-moving star – Procyon in the constellation Canis Minor – were changing over time. Each weaves a somewhat drunken path around a straight line through space, wobbling to one side and then the other. Bessel could find only one plausible explanation for this phenomenon – that the stars were in fact both members of binary systems, being pulled around by companions that remained invisible, despite clearly packing a considerable amount of weight[2].

The case for a companion orbiting Sirius seemed inarguable, but the object itself remained elusive for some years. It was only in 1862 that telescope-maker Alvan Graham Clark, testing what was then the biggest lens in the world*, spotted the star now known as Sirius B. Shining at magnitude 8.44, the lesser star

---

* Clark's father, also named Alvan, was the founder of an eponymous firm that made state-of-the-art glassware for the giant lens-based "refractor" scopes that ruled nineteenth-century astronomy. This particular lens – an 18.5-inch, precisely ground and polished monster – had been intended for the University of Mississippi before the American Civil War stopped play. It was subsequently snapped up for an eyewatering $11,187, becoming the centre-piece of the Chicago Astronomical Society's Dearborn Observatory.

would have been seen long before, had it not been for the dazzling light of its neighbour. Once Clark's report was circulated, many other astronomers found they could spot it now they knew what they were looking for[*].

A few enterprising researchers had already used Bessel's measurements to calculate a 50-year orbital period for the then-undiscovered companion, and this was soon confirmed by plotting the changing position of the companion star itself. Sirius B's eccentric orbit means that as seen from Earth, the separation between the stars varies between a testing 3 seconds of arc (about the size of a 5p piece seen from a mile away) and four times that amount[†]. Today the orbital period has been refined to 50.13 years.

Based on the system's distance and the properties of its orbit, Sirius B is 350 times fainter than the Sun, but has almost exactly the same mass. In 1915, spectroscopy specialist Walter S Adams finally pulled off the tricky task of analysing light from the "Pup Star" alone, using the 60-inch reflector telescope at Mount Wilson Observatory near Pasadena, California. The spectrum showed that, despite its faintness, Sirius B is white-hot – even hotter than its neighbour, and much hotter than the Sun. Its comparative faintness can only be explained if it has less surface area through which to emit light – a lot less[3].

Sirius B, it seemed, was a member of a newly identified class of

---

[*] Decades later, a pair of French anthropologists claimed that the Dogon tribe of Mali had a secret tradition regarding not one, but two, hidden companions of the Dog Star. Unfortunately, this tradition turned out to be so secret that later visitors found no trace of it, but by that time the genie was out of the bottle and an entire cottage industry of books attributing this uncanny knowledge to visiting "ancient astronauts" had spawned.

[†] The separation will be at its widest in 2023.

stars – objects that we now call white dwarfs*. The first of these – part of a triple-star system called 40 Eridani, some 17 light years from Earth – had been identified in 1910 by none other than Henry Norris Russell[†], and had briefly threatened to derail his theories of a neat relationship between the colours and luminosities of stars.

Modern figures actually put Sirius B's temperature a lot higher than that of Sirius A – around 25,000°C, which is hot enough to ensure that much of its radiation is pumped out in invisible ultraviolet. Some fairly straightforward maths reveals that B's diameter is therefore about 0.8% of the Sun's – pretty much the same size as Planet Earth itself.

Fitting slightly more than the Sun's mass into an object the size of the Earth requires matter inside a white dwarf to be packed together to a ridiculous degree. When Ernst Öpik worked out the numbers for 40 Eridani B in 1916 he decried the star's apparent density – 25,000 times that of the Sun – as impossible, but when Arthur Eddington came to tackle the mysteries of stellar structure in the 1920s, he was more open-minded. Using the best estimates available at the time, Eddington found that Sirius B's density must be even greater than that of 40 Erdani B – about 37,500 times more than the Sun. At this density, a sample of Sirius B the size of your little fingertip would weigh over 50 kg, but modern figures show that this was out by a factor of 30 or more – the Pup's true average density is 1.7 tonnes per cubic centimetre.

---

* © Willem Luyten, 1922

† Visiting Pickering at Harvard, Russell asked if Williamina Fleming could dig out the spectral type of "40 Eridani B" and was somewhat alarmed when it turned out to be a hot white A-class star rather than the cool red M he had been expecting. Pickering, however, wisely foresaw that this exception to the rule would lead to great things.

Eddington foresaw that matter in such a tightly packed star would cease to follow the generic "gas laws" that astrophysicists generally applied to modelling stellar innards[*4], but he had no idea at the time of what would replace them. An answer to that question soon appeared, however, from an unexpected quarter – the exciting (and often confusing) new science of quantum physics. Before you run screaming for the exit, I promise to keep this as quick and painless as possible, so [deep breath]…

Quantum physics is the physics of the very small. We're fairly used to the idea that waves of light can behave like particles (usually called photons) but on subatomic levels, tiny particles of matter can also behave like waves. One consequence of this is that many of their properties are "quantized" – like notes from a vibrating string, they can only take on certain distinct and separate values. An electron orbiting inside an atom, for instance, has an energy defined by the size and shape of its orbit: it can have energy X or energy Y, but never anything between these values, just as a violin string can only create sustained notes by vibrating at certain frequencies[†]

In 1925, the brilliant Austrian physicist Wolfgang Pauli discovered the famous "exclusion principle" that bears his name. This basically says that no two particles in a system[‡] can share the same set of "quantum numbers" – they're forced to arrange themselves in ways that avoid two particles having a completely

---

[*] School physics lesson staples defining relationships between the pressure, volume and temperature of gases.

[†] String theory, one potential "Grand Unifying Theory" for particle physics, takes this metaphor pretty much literally.

[‡] Notoriously flexible physics-speak for a self-contained region containing whatever you're studying – anything from an individual atom to the whole Universe, really.

identical set of quantized properties. This explains why electrons fill up the various possible orbits around an atomic nucleus in the way they do, and as a result essentially describes the structure of matter itself.

[... and relax.]

A year after Pauli's breakthrough, Ralph Fowler, a Cambridge physicist who had been awarded an OBE after World War I for his work on anti-aircraft ballistics, applied the new theory to the interior of the white dwarf stars. The basic idea behind Fowler's somewhat daunting paper[5] was to model these superdense stars as a single giant quantum system. Since atoms are mostly empty space*, and temperatures above a few thousand degrees can break nuclei and their orbiting electrons apart, the interior of a typical star is essentially a particle soup of electrons and nuclei, only saved from collapse by the outward pressure of escaping energy. Fowler realised that if you cut off the energy source, then the star would fall inward under its own weight until the exclusion principle, acting to keep electrons in different quantum states, exerted a pressure of its own to slow and halt the collapse. One curious consequence of this situation is that higher-mass white dwarfs like Sirius B are smaller, and have hotter surfaces, than those with relatively low mass like 40 Eridani B.

So where do white dwarfs come from? Fowler's work showed that a star can only become a white dwarf when its internal energy supply is switched off, so it shouldn't come as too much of a surprise to learn that they're the final stage in the life cycle of many stars and foretell the ultimate fate of our own Sun. In the previous chapter, we saw how a star swells to become a red

---

* Imagine a golf ball on the centre spot of a football pitch, orbited by a grain of rice whizzing around in the nosebleed section.

giant as it runs out of hydrogen fuel in its core, and eventually becomes unstable once it has plundered the helium waste products of its main-sequence lifetime. The end result is a star enriched with the products of helium fusion (principally carbon and oxygen) but affected by increasingly violent pulsations as two onion-skin shells of hydrogen and helium fusion move out through the layers of the star, around the burnt-out core. Eventually, these pulsations become so great that the star loses its grip on its outermost layers, puffing off a series of cool cosmic smoke rings over tens of thousands of years. Removing these outer layers means that the star generates less heat and pressure during its phases of contraction, so the fusion in the shells burns less fiercely.

The last gasp end of a red giant sees its remaining layers lost in two directions at once – either collapsing inward due to the gravity of the core (which has been growing denser ever since it stopped generating energy) or blown outward by the remaining force of radiation as the fusion shells falter and die. Material falling onto the core gives white dwarfs a characteristic spectrum rich in carbon and oxygen, while ultraviolet radiation now escapes from an exposed surface heated to tens or even hundreds of thousands of degrees. As these UV rays pass through the expelled outer shells, they cause them to fluoresce (in a similar way to the gas around hot newborn stars in emission nebulae). The result is a short-lived glowing gas cloud called a planetary nebula[*], with a developing white dwarf star at its heart. Depending on conditions around the dying star, planetary nebulae can range from delicate

---

[*] Nothing to do with planets themselves, of course – late-eighteenth century astronomers simply noticed the resemblance between their typically pale disc shape and their view of planets such as Jupiter.

rings and spherical bubbles, to twin-lobed hourglasses (pinched at the middle by a ring of denser, slower-moving material) and even twisted cosmic seashells shaped by the influence of other nearby stars. Often, overlapping veils of gas are picked out in different colours as the elements within them fluoresce at different wavelengths, creating some of the most beautiful objects in the entire night sky.

Finding a cosmic smoke ring

Planetary nebulae may be beautiful, but they're also challenging to spot. The best bet for stargazing beginners is one called the Dumbbell Nebula lurking in the otherwise insignificant constellation of Vulpecula, the Little Fox. Although there's nothing particularly fox-like about this part of the sky, it's still easy to locate because it's runs along the south side of cross-shaped Cygnus, the Swan.

The spot we're interested in, however, is best found with reference to the constellation on Vulpecula's southwestern flank, Sagitta (the Arrow). This narrow group of four stars really does look like a bit like a celestial cursor, and if you scan north (towards Cygnus) from the bright star on the arrow's tip, you should come to a fuzzy patch of light about one-third the size of a Full Moon (the distance is less than the length of Sagitta's arrow itself).

Through binoculars or a small telescope, the Dumbbell lives up to its name with two bright lobes of gas extending from its centre. Long-exposure snaps that soak up more light show that these lobes are just the brightest parts of an

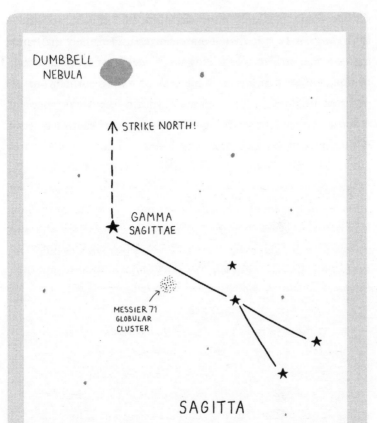

oval gas bubble, thrown out by a central star of magnitude 13.5 that is in the process of collapsing to a white dwarf. Based on the nebula's distance of more than 1,200 light years, this bubble is almost exactly a light year wide. By measuring its speed of expansion, researchers have further figured out that it began to form about 10,000 years ago, making it a cosmic youngster compared to most other objects in the sky.

While Sirius B seems puny alongside its brilliant neighbour today, we can be sure that at one stage it was definitely the senior partner. Just as Sirius A is using up its fuel at an accelerated rate compared to the Sun, so its sibling B must have burned through it faster still, living fast in a blaze of brilliance and dying young as a spectacular red giant. Best guesses suggest this happened about 120 million years ago, when the Sirius system as a whole was about 115 million years old*. In order to reach such an advanced stage of stellar evolution so quickly, the Pup must have started out with about five times the mass of the Sun. Sirius A, meanwhile, still has another few hundred million years to go before it in turn swells to a red giant, sheds its layers in a planetary nebula, and ultimately diminishes to become a white dwarf in its own right.

---

* This, along with the distinct lack of any expelled gas lingering in the area, blows a hole in the otherwise-enticing theory that Sirius B was a red giant in historical times. Certainly, several Roman writers, including the philosopher Seneca and Ptolemy himself, described Sirius as red, but the solution to this puzzle probably lies more in the realms of human perception than astrophysics.

14 – RS Ophiuchi

*A star that occasionally goes bang,
and will one day go boom*

---

*O*ur next star is something of an oddity – most of the time if you go looking for it with anything less than a half-decent telescope, you'll simply find an empty spot in space. But catch it at the right moment, or keep an eye on the Twitter feed for the American Association of Variable Star Observers*, and you can see the brief flare-up of one of space's most impressive events.

RS Ophiuchi (the letters come from the same makeshift approach to cataloguing bright variables that gave us T Tauri) lies within the expansive bounds of Ophiuchus, the Serpent Bearer. This faint, somewhat ungainly constellation represents a giant wrestling a snake (neighbouring Serpens†), and has been associated with legendary figures from the Greek god Apollo and the ill-fated Trojan priest Laocoön, to Asclepius, the mythical proto-physician whose snake-entwined staff is still a common medical symbol today. The main stars in the pattern look like

---

* @AAVSO

† Serpens is actually divided into Serpens Caput (the head) west of Ophiuchus, and Serpens Cauda (the tail) to his east, making it the only split constellation in the sky.

the end of a hip-roofed barn. For those of us in the northern hemisphere, they glide across the southern evening skies from around May to October, while early risers can spot them from February onwards. Southern-hemisphere skywatchers can see Ophiuchus crossing northern skies in the same months – and although the constellation is upside down, it's perhaps a little easier to see its pattern by looking for the shape of a large shield.

The Serpent Bearer's stars aren't exactly the brightest, but your best bet for finding our target star (or at least its location) is to scan with binoculars or a small telescope down the eastern flank of the main pattern (marked by yellowish Cebalrai at the northern end and pure white Sabik at the southern). Slightly north of halfway between them, and a little to the east, you should spot a fuzzy

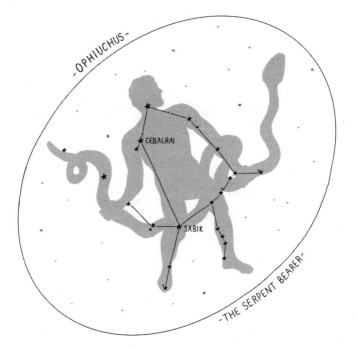

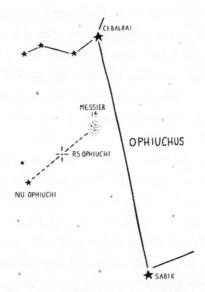

ball – a distant "globular" star cluster called Messier 14*. Now scan southeast along a line between M14 and Nu Ophiuchi at the border with Serpens – RS is almost exactly halfway along.

As we've already said, though, it's wise not to get your hopes up – normally this star loiters somewhere around magnitude 11.5 (though its quiet periods vary considerably). Once every couple of decades, however, it becomes a fairly obvious naked-eye star for a couple of months, with outbursts that usually peak at around magnitude 3 or 4. RS Ophiuchi is the poster child for a class of stars that today's professional astronomers call "cataclysmic variables". Most amateurs, however, still prefer to use a term with a couple of millennia of history behind it: novae.

It doesn't take a classical education to guess that "nova" comes from the Latin for new – in this case, a new star in the sky. The

---

* An early entry in French comet-hunter Charles Messier's 1771 list of fuzzy objects that later turned out to be star clusters, nebulae and galaxies.

earliest nova on record burst into visibility around 134 BCE in the constellation of Scorpius and was spotted by the astronomer Hipparchus, then living on Rhodes. According to Pliny the Elder (writing a couple of centuries later) the sight of an unexpected new light in the sky, behaving very differently from a comet, inspired Hipparchus to compile the ancient world's first detailed star catalogue*.

Chinese astronomers were probably recording novae for centuries before Hipparchus, but their earliest records were lost thanks to the revolutionary policies of Qin Shi Huang, the all-powerful First Emperor, who implemented widespread book-burnings in the third century BCE – and had scholars who attempted to preserve their knowledge buried alive. As China recovered from this bloody excess, record-keeping resumed and most of our knowledge of pre-Renaissance novae come from records of *kèxīng* or "guest stars". Aside from a few vague allusions, however, records from the classical and European world are mostly silent on the subject of novae until the Renaissance – perhaps because scholars there remained in thrall to Aristotle's idea that the heavens beyond the Moon were perfect and unchanging (the same reasoning meant that comets, which clearly *did* change, were assumed to be phenomena of the upper atmosphere).

When a brilliant new star appeared in Cassiopeia in November 1572, however, the flaws in the increasingly stodgy Aristotelian view became obvious to anyone who didn't have their ideological

---

* Sadly lost, along with most of Hipparchus's other work. Historians argue over quite how much Ptolemy plagiarised it for his own *Almagest* in the mid-second century, and there's a lovely (though debatable) theory that Hipparchus's chart is preserved on the Farnese Atlas, a second-century statue of the eponymous Greek Titan shouldering the weight of the celestial sphere.

blinkers firmly fixed. Recorded in detail by the last generation of pre-telescopic stargazers, this star (technically a supernova rather than a nova, though that distinction wouldn't be recognized for centuries) paved the way for wider debate about the hypothesis of a Sun-centred Universe, an idea that had begun to take flight in Copernicus's deathbed manifesto of 1543*.

For astronomers of the new enlightened age, novae were an intriguing but frustrating subject for study. Flaring brilliantly into life over the course of perhaps a few days, they took months to fade from view, but remained small, star-like points of light throughout this cycle. Various theories were put forward to explain them – perhaps they really were new stars (though in that case why didn't they stick around?) or followed orbits that brought them rapidly towards Earth before a slow retreat? One of the most prescient ideas, however, came from master physicist, astronomer and occasional alchemist Isaac Newton in 1713. He suggested that a nova might be a burnt-out, faded star, briefly flaring into life thanks to the impact of comets on its surface.

The first concrete breakthrough in the understanding of novae, however, came when everyone's favourite husband-and-wife spectroscopy pioneers, William and Margaret Huggins, captured the spectrum of a nova that went off in the constellation of Auriga in 1891[1]. The star's light showed both bright emission lines and dark absorption lines, shifted from their usual positions by Doppler effects both towards and away from Earth. The whole thing was best interpreted as a vast cloud of gas expanding at

---

* When a brilliant comet turned up in 1577, and Danish master stargazer Tycho Brahe showed definitively that it was moving through the space beyond the Moon, that should have just about wrapped it up for Aristotle – but rather like Monty Python's Black Knight, his theory continued to hop around, insisting "'twas but a scratch" and picking fights with Galileo and others until well into the seventeenth century.

speeds of hundreds of kilometres per second and varying wildly in temperature from one part to another – in other words this nova, at least, was the site of an explosion.

RS Ophiuchi itself was discovered a few years later – not at the sharp end of a telescope but in the photographic archives of Harvard College Observatory. Here the indefatigable housemaid-turned-astronomer Williamina Fleming spotted tell-tale signs in 1899 of nova-like activity in the spectrum of an otherwise unremarkable star. Looking back over various plates that had previously captured the star's spectrum, she found that it had changed substantially since 1894. Fleming's fellow "Harvard computer" Annie Jump Cannon, sifted through photographs of the region, and by charting the star's changing brightness over time (a pattern astronomers call its "light curve") and comparing it to other novae, she found that it must have peaked in 1898[2].

While novae typically spike in brightness over just a few days, they take weeks or longer to decline, and are often prone to unpredictable smaller variations afterwards. Some (but not all) even go through a brief revival in brightness before their fading resumes. This somewhat unpredictable behaviour has long made them a favourite for amateur stargazers looking to do some actual science. The professionals can't be everywhere at once, but thanks to organisations like the AAVSO (founded by William Tyler Olcott in 1911 and actually open to members from around the world), and rapid-response alert systems such as the Central Bureau for Astronomical Telegrams[*], hobbyists have long been able to do what we'd today call "citizen science". This is one reason why, when RS Ophiuchi began to brighten dramatically

---

[*] Operating these days out of Harvard – and despite the name they're firmly in the digital age.

once again in 1933, its eruption was quickly spotted[*] and news circulated around the world.

Thanks to the work being done at Harvard, astronomers had realised as early as 1902 that stars could become novae more than once[†], but RS Ophiuchi is one of only two such stars that can reach naked-eye brightness. Since 1933, RS has erupted in 1945, 1958, 1967, 1985 and 2006, fully earning its designation as a "recurrent nova". Its rival in the peak brightness stakes, T Coronae Borealis (in the pretty circlet of stars known as the Northern Crown), is a slacker by comparison, with outbursts recorded only in 1866 and 1946 so far. Thanks to its brightness and frequency of eruptions, RS has probably been studied more than any other nova system, and has played a vital role in building up our picture of how these strange stars work.

★ ★ ★

So what exactly is going on when a star turns nova? The explanation turns out to be not a million miles removed from Newton's "reinvigoration by comets" theory. A key breakthrough came in 1955, when young Caltech astronomer called Merle F. Walker took a look at what was left of DQ Herculis, a nova that had erupted two decades before[3]. Walker used electronic photometers attached to the giant 100-inch reflector telescope at Mount Wilson Observatory in California, to capture tiny fluctuations in the light from the 15th-magnitude star, and found a clear pattern in the form of regular, abrupt dips in brightness.

---

[*] First off the mark was Eppe Loreta, an Italian librarian tracking variable stars from his private observatory in Bologna.

[†] T Pyxidis, a 1901 nova discovered after the fact on photographic plates taken at Harvard, inspired a trawl through the archives that revealed an earlier 1890 eruption.

The signature was unmistakeable – DQ was an eclipsing binary system somewhat similar to Algol, in which a faint star passes in front of a brighter one and partially blocks its light once every orbit. It soon became clear that there were two key differences, however: firstly, DQ's dips were *tiny* compared to those of most other eclipsing binaries, which suggested that the object doing the eclipsing was pretty small; and secondly, the eclipses repeated very rapidly, every 4 hours 39 minutes. Because the speed at which stars orbit each other is related to the force of gravity between them, and the visible star was a lightweight red dwarf, this suggested that the unseen companion must have a pretty impressive mass.

Intrigued by this discovery, Robert Kraft, a young researcher working at Mount Wilson, began a systematic hunt for signs that nova systems were close binaries. Walker had got lucky with DQ Herculis, which happened to be aligned at the right angle to create eclipses, but even if other nova systems were binary, the vast majority of them would offer no such giveaways. Kraft set out instead to measure "wobbles" in their radial velocity – signs in the stars' spectral lines that the system's main source of light was being pulled back and forth by an unseen companion (an early application of the method today's planet detectives use to find alien worlds around stars like Helvetios).

Shuttling back and forth between US observatories as a young, untenured academic, it took Kraft some time to gather his data but by the early 1960s, he had assembled a dossier of evidence that a wide range of nova and nova-like systems – including one-offs, recurrent novae and "dwarf novae" (stars prone to smaller but more frequent outbursts) – were in fact binaries[4]. And on top of this, they were binaries with some very specific properties: the star that provided all the light during the system's dull "quiescent" phase was being pulled around by a heavyweight, invisible companion.

In the vast majority of cases, there's only one candidate that fits the description of Kraft's invisible companion star – a white dwarf. But what role could a burnt-out stellar core play in triggering a nova outburst? The answer to that finally became clear in 1971 thanks to a pioneering application of computer modelling by Sumner Starrfield at Yale University Observatory.

By this point, inspired by Kraft's work, astronomers had been subjecting "old novae" (stars with a nova outburst in their past) to intensive study. They found that the normal stars in each system could vary considerably – from red dwarfs through brighter main-sequence stars all the way up to red giants – but there was a link between the type of star and the orbital period: novae involving small stars had extremely short orbital periods, while those involving giants could be much longer (RS Ophiuchi itself, for instance, involves a red giant and a white dwarf orbiting each other in around 454 days[5]). Meanwhile, other studies were finding evidence for hydrogen being lost from the main star and floating around the system before settling onto the white dwarf, forming a fresh "atmosphere" around the exhausted stellar core.

Remember on our visit to Algol where we saw how the paradox of the two stars' ages could only be solved if one had over-spilled its gravitational boundaries and transferred material onto the other? It soon became clear that a similar process is going on with novae: the companion is always larger than the safe "Roche lobe" where it can be sure of holding onto its material, and is therefore leaking gas from its outer layers that gets swept up by the intense gravity of the dense white dwarf*. As this hydrogen approaches the dwarf, it clumps, jostles together and heats up, eventually settling into a flattened "accretion disc" (a bit like the rings around Saturn) through which material gradually spirals down onto the white dwarf itself.

But just which element of this complex stellar juggling act produces the actual nova explosion – and how? Different researchers had claimed evidence that they came from the dwarf, the disc or the companion, and it was only Starrfield's modelling that put the solution beyond doubt[6].

The numbers showed that nova explosions result from a violent outbreak of nuclear fusion, known as a thermal runaway, on the surface of the white dwarf. This is only possible because of the strange properties of its compressed matter the greater the dwarf's mass and the more matter it contains, the smaller it is. So when hydrogen falls onto the surface, it gets heated by the incandescent material already there, but is also compressed so that it cannot simply escape by boiling away into space. As more and more hydrogen is siphoned from one star to the other, the dwarf builds up a layered atmosphere whose innermost regions become ever hotter and more densely packed.

You might be able to guess what happens next – the densely packed hydrogen eventually gets so hot that it triggers a new wave of nuclear fusion. What's more, the heat and pressure are so great that these aren't the moderate proton-proton chain reactions seen in the Sun – instead, fusion can follow the far more rapid CNO cycle (which we saw rising to prominence in Sirius). A runaway

---

* The range over which this can happen depends not just on the mass of the white dwarf, but also on the size and density of the companion star – hence red giants are vulnerable to Grand Theft Hydrogen over distances where a normal star would be safe.

reaction rips through the hydrogen envelope, and a process that is normally concealed deep inside the deeply buried cores of stars, is exposed to light up the Universe.

At a rough estimate, the average nova eruption is triggered when a white dwarf accumulates about $1/10,000^{th}$ of a solar mass of material on its surface. Only a fraction of this superdense atmosphere (perhaps 5 per cent) gets converted from hydrogen into helium before the pressure buildup blasts the rest away into space, but for a few days, the rate of fusion reactions is so fast that a nova can pump out up to 100,000 times more energy than the Sun. Astronomers reckon that this must happen about 50 times per year in our galaxy on average[7], but even with the most powerful telescopes, the discovery rate typically averages less than a dozen (blame all the intervening clouds of stars, gas and dust getting in the way – the Milky Way is a mucky place, and well overdue for a deep clean).

Despite the violence of their outbursts, however, novae are rarely powerful enough to disrupt the fundamentals of the systems that create them[*]. After a while, the stars recover, the link between the white dwarf and its neighbour is re-established, and "normal service" is resumed. Only a dozen or so recurrent novae have so far been caught in the act of repeated eruptions, but studies of other "post-nova" star systems have also confirmed fresh layers of hydrogen piling onto the dwarf star component. Every nova, it seems, is a recurrent nova waiting to happen – it just might take a few millennia.

---

[*] They do, however, create an expanding halo of gas known as a nova remnant, moving at thousands of kilometres per second. Following RS's 2006 outburst, radio astronomers were able to pin down the nova's distance by comparing the speed of the remnant's expansion (established by measuring its spectrum) to its growing diameter in the sky. They concluded that it is about 4,600 light years from Earth.

Both the period in which a nova repeats and the strength of its outbursts depend on several factors, and RS Ophiuchi's unique set of attributes make it more prone to reasonably frequent, bright outbursts than any other star in the sky. While the stars are quite widely separated (about 10 per cent further apart than the Earth and Sun), the fact that the companion is a bloated red giant means that its outer layers are within easy reach of the hungry white dwarf. The dwarf's gravity is also particularly strong because it's just about as heavy as a star of its kind can get, at about 1.4 times the mass of the Sun. Stronger gravity not only means a more rapid accumulation of gas from its surroundings, but a more vice-like grip on its atmosphere during outbursts, enabling fusion to continue for longer and the nova to shine more brightly before the shredded upper layers of the atmosphere finally escape and bring the eruption to an end.

If you're wondering why white dwarfs never get much above 1.4 times the Sun's mass, then the simple answer is that stars with heavier cores collapse into even smaller, denser objects, like neutron stars and sometimes even black holes. We'll be exploring both of these objects in later chapters, but it's worth discovering the lives of the most massive stars before we focus on their deaths.

In the meantime, though, there's one final point to make about RS Ophiuchi. As it teeters on the upper mass limit for a white dwarf, each cycle of accumulation, outburst and dissipation leaves a little more material on its surface, pushing it another step closer to a cliff edge where pressure between electron particles (the internal support mechanisms for all white dwarfs), can no longer withstand the inward pull of gravity. One day, probably within the next 100,000 years or so, RS Ophiuchi will plunge off the precipice in a sudden, dramatic and total collapse. The energy released in this event, called a Type 1a supernova, puts even the brightest nova eruptions in the shade, reliably peaking at a staggering 5 *billion* times the output of the Sun. We'll have a lot more to say about this process when we come to our very last star.

To put it another way, despite a distance of thousands of light years, RS Ophiuchi will one day become a beacon in Earth's skies, shining 600 times more brightly than Sirius does today. If only for that reason, it's certainly worth keeping an eye on.

15 – BETELGEUSE

The biggest stars in the sky and
how to measure them

*

eetlejuice, Bettelgoose, Betelgeuse – this hard-to-pronounce star is as famous today for inspiring Tim Burton's cult comedy horror movie as it is for being one of the brightest stars in the sky. But once we've settled the issue of its pronunciation (say the first "e" like "feather", and the "g" like "badger", say the experts), it turns out that Betelgeuse is not just easy to spot – it's played a key role in improving our understanding of the biggest stars of all, and inspired our methods for observing fine detail in the cosmos.

And Betelgeuse really is one of the easiest stars in the sky to track down. Look for it on the shoulder of Orion, the Hunter[*], as he makes his graceful arc across evening skies between November and March (or in morning skies from July if you're an early riser). North of the equator, Orion's an obvious human figure with a

---

[*] Betelgeuse's name actually comes from the Arabic *yad al-Jauza*, meaning "hand of the central one". Early Arabic astronomers depicted the central one in question as a fearsome female warrior, and some think the *al-Jauza* in question might be a pre-Islamic deity descended from ancient Mesopotamian sex-and-war goddess Ishtar.

chain of three stars (his famous belt) pulling in his waist. Facing the charging bull Taurus (another pattern that you really can't miss), he stands with legs heroically akimbo in a power stance that would make a politician blush. The bright blue-white star Rigel marks his foremost knee, while Betelgeuse sits at the opposite shoulder of an upper body that would put most weightlifters to shame. Get away from the city lights, and you'll see faint chains of stars making up a raised shield, a club, and of course the sword dangling from his belt (home to the famous Great Orion Nebula and Trapezium that we've already encountered).

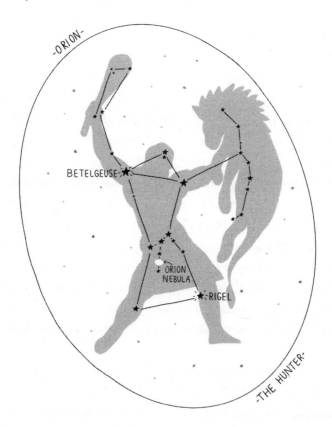

For northern hemisphere observers, Orion sits neatly at the centre of a great tableau of constellations, with the rampant bull ahead of him, two hunting dogs Canis Major and Minor behind (both marked by bright white stars, including Sirius, the brightest of all), and a terrified bunny rabbit (Lepus, the Hare) fleeing the scene at his feet. South of the equator you'll have to interpret this epic upside down, with Betelgeuse below and to the right of Orion's belt.

But whatever angle you're looking from, Betelgeuse is unmistakable – not just because of its brightness, but also due to its obvious red colour. It's noticeably ruddier than Aldebaran, the eye of nearby Taurus, and so distinctive that nineteenth century Italian astronomer Angelo Secchi, who came up with one of the first classification schemes for stellar colours, used it as his keystone for orange-red stars. In modern terms, Betelgeuse's colour puts it in spectral class M1 or M2 – suggesting a surface temperature of about 3,300 °C, which is about as cool as stars get.

As we've already seen, cool red stars can be at one of two extremes. They're either feeble red dwarfs like Proxima Centauri, far fainter than the Sun and all but invisible unless they're on our cosmic doorstep, or brilliant red giants like Mira, cool only because they've swollen to enormous size during the final stages of their lives. As one of the brightest stars in the sky, it should come as little surprise that Betelgeuse falls into the latter category – but in fact, when it comes to size, this star in a completely different league.

Astronomers realised that Betelgeuse must be something of a whopper once Hertzsprung, Russell and others began to piece together the relationship between star brightness, colour and size early last century. By this time, the first attempts to measure its parallax – the tell-tale annual shift in perspective that could betray its true distance – had already been made. Sir David Gill, working at the Cape Observatory in South Africa, found no detectable movement, but William Lewis Elkin came up with a figure of 0.024" (24 thousandths of a second of arc, or just over 1/200,000[th] of a degree), indicating a distance of at least 135 light years[1]. Vague though these measurements were, they clearly showed that Betelgeuse was *much* more luminous than the Sun – and if a star so luminous was also cool and red then it must have a *very* big surface area for all that energy to escape through. If cool, orange Aldebaran was a giant, then it stood to reason that the more distant, ruddy Betelgeuse must be even more gigantic still.

In an address to the British Association in 1920[2], the always-perceptive Arthur Eddington put all this together to predict that Betelgeuse was the star with the largest angular diameter visible from Earth, offering the best chance of actually seeing another star as a measureable disc rather than an infinitesimal point of light.

The principle obstacle to actually measuring this diameter, it became clear, was not the magnifying or resolving power of contemporary telescopes, but the blurring effect of the atmosphere (which inevitably smears out light, causing stars to twinkle and leaving images less than pin-sharp on even the clearest of nights). However, Eddington's prediction came just at the right moment, for just a few months later, scientists at California's Mount Wilson Observatory were able to measure (if not directly observe) Betelgeuse's diameter with a revolutionary new instrument – the very first large astronomical interferometer.

A *what?* Well interferometry is tricky to explain, not least because it's a multi-faceted technique that can be implemented in many ways and has countless applications across different areas of science. The basic idea, however, is that you can get extra information about a light source by combining beams that have travelled along different "optical paths", and seeing how they interact with each other*.

Because light consists of fast-rippling waves of energy, combining two beams creates interference, as the waves reinforce each other in some places and cancel out in others. The result is a series of "interference fringes" whose pattern is hypersensitive to the length of the paths along which the two beams have travelled. You can rig up an interferometer apparatus to measure such fine differences in a variety of ways. The Mount Wilson interferometer, for instance, was designed to create two images alongside each other that could be seen through an eyepiece: a control image that would *always* show fringes if things were working properly, and another in which the fringes would vanish if the star had an "angular diameter" over a certain size.

With the 6.1-metre-wide steel scaffold of the interferometer bolted onto its front end, the observatory's state-of-the-art 100-inch Hooker Telescope was temporarily relegated to a mere support act – quite literally, since its main job was to point the interferometer in the right direction. Two angled mirrors on opposite sides of the frame were fixed in place, while another pair at right-angles could be slid back and forth, adjusting the distance between them and, with that, the instrument's sensitivity.

The entire apparatus was the brainchild of renowned physicist Albert Michelson† and the somewhat less-renowned astronomer

---

* You can also turn the technique on its head and learn about other stuff by splitting up a beam from a light source you already understand (a laser, say), and letting its two halves have slightly different adventures of their own before recombining them – that's the principle used to detect gravitational waves, for instance.

† Michelson had already become the first US Nobel Prize-winner for other precision optical measurements – including the experiment that got Einstein wondering about that whole speed-of-light business. Rather unexpectedly, he's also the subject of an episode of 1960s TV horse opera *Bonanza*.

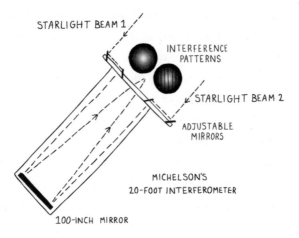

Francis G. Pease. On 13 December 1920, they took a first look at several stars with the moveable mirrors set at 229 cm apart. Their first two targets produced fringes in both images, but as they'd hoped, with Betelgeuse the fringes on the test image vanished: even if they couldn't distinguish it by eye, this star's image was a disc rather than an infinitesimally small point of light.

Michelson and Pease crunched the numbers and found that Betelgeuse must have an angular diameter of 0.047". Combined with the most recent parallax measurements of the time, this showed that the star was an astounding 386 million kilometres wide, making it 278 times bigger than the Sun and just slightly smaller than the orbit of Mars[3].

We'll come back to the question of its *precise* diameter later, but this initial measurement was certainly enough to confirm Betelgeuse as the first and nearest in a new class of stars, now known as supergiants. Pretty soon, many more of these stars were identified (though more often through analysis of their spectra than direct measurement of their size). They shared incredible luminosities, and sizes, but were spread across a whole range of colours with red supergiants the largest of all (take a look back at page 68 if you want to see how they fit into the H-R diagram of stellar properties). Astronomers differed on where these behemoths sat in the general pattern of stellar evolution, and it wasn't until the 1940s that the confusion began to clear.

We covered our understanding of stellar death throes in some detail when we looked at Mira, but there we were mostly concerned with stars of roughly solar-mass. And we'll hold off until the next chapter to take a more detailed look at the internal workings of real monster stars like 12-solar mass Betelgeuse and the even heavier Eta Carinae, both of which are destined to end their lives in cataclysmic supernova explosions. For now, it's enough to know that the greater mass of these monsters means their internal engines run at much higher temperature and pressures, burning through much more fuel far more quickly. That makes them shine even more brilliantly than normal red giants, and the pressure of radiation from within inflates their outer envelope of hydrogen gas whose glowing surface defines the star's visible photosphere, to bilions of kilometres in diameter.

The tenuous nature of a red supergiant's outer layers, however, makes it extremely hard to work out *exactly* how big it is. While gas in a star like the Sun turns from opaque to transparent across a fairly narrow layer and so appears to form a sharp-edged luminous surface stars like Betelgeuse are much more blurry around the edges. Light can break free at different depths depending on its wavelength, resulting in varying amounts of "limb darkening", a fading effect around the edge of a star's disc.

Michelson and Pease took a rough guess that Betelgeuse's true angular diameter, with these effects taken into account, was probably around 0.055 seconds of arc In 2000, a team used the Infrared Spatial Interferometer at Mount Wilson* to measure Betelgeuse at mid-infrared wavelengths (where limb darkening is smallest and the measured diameter should be at its maximum), and landed on that exact figure. A few years later, however, measurements in the near-infrared delivered a significantly smaller result of 0.043". The team behind this measurement suggested that the difference is due to a cool layer of gas above the actual photosphere, which glows in visible and mid-infrared light, but is otherwise transparent.[4]

So if we're trying to estimate the true diameter of Betelgeuse at its photosphere, then the smaller measurement is the more accurate one. Coupled to the latest and most accurate parallax measurement of 0.0045" (which situates the star around 720 light years away), it puts Betelgeuse's diameter at 1.3 *billion* kilometres – comfortably swallowing up our solar system's asteroid belt but a little smaller than the orbit of Jupiter.

★ ★ ★

Measuring the diameter of stars using clever tricks is all very well, but creating actual images and seeing details on their surface is another challenge entirely. Fortunately, in the past few decades another form of interferometry has finally allowed ground-based telescopes to overcome the limitations of the atmosphere and create images in fine detail.

The foundations for this technique were laid as early as 1963 when French telescope engineer Jean Texereau published his

---

* A rather different instrument from the 1920 interferometer, this array of three 1.65-metre infrared 'scopes cribs a technique called aperture synthesis from radio astronomy to achieve the resolving power of a 70-metre mirror – for a little more on this, see the chapter on 3C 273.

analysis of what's really going on as light descends through Earth's turbulent atmosphere[5]. He pointed out that if you could look at the messy image of a star from instant to instant, you would see a single speckle of light dancing back and forth around the star's true position, deflected from its original path by the lens-like effects of the rippling atmosphere.

A few years later another French astronomer, Antoine Labeyrie, showed that you could (in theory) recover an undistorted image from the moving speckle[6]. A fiendish piece of mathematical trickery called a Fourier transform* can break down the various moves in the speckle's dance into periodic shifts of different frequencies. Once the Fourier transform has revealed the hidden order among the speckles, recombining them into a pin-sharp image is mostly a matter of applied computing power.

Betelgeuse was an obvious guinea pic for the new technique, and was used for various tests in the next few years. It wasn't until 1989, however, that processing reached a stage where something close to a coherent image was achieved[7], using the William Herschel Telescope on La Palma in the Canary Isles.

There are various ways of putting speckle interferometry into practice[†], but David Buscher and his fellow Betelgeuse-botherers using a method called "aperture masking". This involved blocking the front of the 4.2-metre telescope with a plate containing just a few tiny and strategically placed holes. The pinprick beams of

---

* Don't ask. Really, don't ask unless you're willing to put in a couple of years of degree-level maths – I still have nightmares.

† In these days of responsive CCDs and a supercomputer in every pocket, amateur astronomers use a simpler method called "shift and add", taking video clips of bright objects such as planets, which are then selected and realigned frame-by-frame to make a sharp final image.

light that passed through were then recombined, and the resulting interference pattern used to reconstruct the image.

By the time it had been churned through a late-80s vintage computer, the final image of Betelgeuse looked more like a contour map than a photo – but it was still enough to reveal the supergiant's distinctly asymmetric shape, with an off-centre bright spot. At the time, some wondered if this could be a hitherto unsuspected companion star passing in front of Betelgeuse, but most accepted the more likely explanation that they were looking at the top of a vast, bright spot of rising gas, hotter than the rest of the star's outer atmosphere.

This was confirmed a few years later when NASA's Hubble Space Telescope (using a rather more direct solution to the distortion problem), snapped Betelgeuse from high above Earth's atmosphere in 1996. The first truly direct image of a star's surface, taken by the simple "point and shoot" method rather than inter-ferometry, revealed a vast extended atmosphere and a large "hotspot", 2,000°C above the average surface temperature in the photosphere.

In the past two decades, new images of Betelgeuse have come thick and fast. Earth-based interferometry and computing have improved in leaps and bounds – to a point where they can reconstruct colour pictures of Betelgeuse and a select handful of other stars. They've shown bright hotspots coming and going from the surface in a matter of weeks, and have revealed vast clouds of gas extending for a trillion kilometres around Betelgeuse – a sure sign that the star is prone to shedding huge clouds of gas as waves of radiation from the interior overcome the tenuous grip of gravity on its distant outer atmosphere.

★ ★ ★

With a tenuous outer atmosphere and multiple layers of internal fusion, it's no surprise that Betelgeuse is a wee bit unpredictable, and the first person to bring this to widespread attention was John Herschel in 1840*. Compared to the near-constant light of nearby Rigel, he noted that Betelgeuse was usually fainter (dipping to a minimum magnitude of around 1.2) but sometimes a little brighter – even hitting magnitude 0.1[8]. Perhaps Johann Bayer caught it on a good day when he gave Betelgeuse top billing in the Greek alphabet, labelling it as Alpha Orionis?

Almost two centuries of observations since have shown Betelgeuse alternating between quiet periods in which it's relatively dim, and more active spells when it fluctuates and reaches peak brightness. Astronomers class it as a semiregular variable – its short-term pulsations are probably due to a similar balancing act between different fusion shells to that we saw inside the more dramatic variable Mira. The longer-term ones remain a puzzle, but may be linked to whatever process deep inside the star produces the vast hotspots.

Just as with Mira, Betelgeuse's changing brightness is accompanied by changes in its size, and measurements of the Doppler shift in light from its atmosphere show periodic expansions and contractions. Beyond this, however, it's sadly difficult to say much more – Betelgeuse's tendency to deliver contradictory results that change with wavelength and observing tech extends to properties other than its diameter. A 2009 study that observed Betelgeuse consistently over the course of 16 years, for example, concluded that the star had

---

* Widespread *Western* attention, that is – there's a good chance that aboriginal Australians noticed Betelgeuse's changeable nature centuries or even millennia ago and recorded it in their oral tradition of lustful hunter Nyeeruna and his club, which was said to periodically fill and empty with "fire magic" as he faced off against Kambugudha (the Hyades), defensive elder sibling of the beautiful Pleiades sisters.

been shrinking at a dramatic rate and was now 15% smaller than it had been in 1993, but then two years later, two of the same authors decided a more likely explanation was the changing transparency of a gas shell above the star's actual surface[9].

For this reason, it's perhaps wise to take reports of dramatic changes with a pinch of salt. For instance, the post-Christmas doldrums of 2019 were enlivened by reports of a dramatic dimming from our favourite supergiant, knocking it out of the top 20 brightest stars in the sky for the first time. Reporters enthusiastically speculated that the Big B might be entering its final death throes and nearing supernova, ignoring the evidence that it's still got a few phases of nuclear fusion (lasting perhaps another 100,000 years) before it reaches that point. What actually happened, wiser heads agree, was a coincidental overlap between the regular long- and short-period pulsations, leading to an extra-deep dip in light output. On top of this, the star seems to have suffered from a particularly virulent outbreak of dark starspots in its outer layers.

While Betelgeuse's many veils may repeatedly confound our attempts to get to grips with it, the star's huge size, relative proximity to Earth and unusual place in the great scheme of stellar evolution make it hard to resist. Amateurs can still do their bit in measuring its changing brightness, while for professionals, it's been an inspiration and a testbed for the development of high-precision observing techniques – but for all that we may have learned about it, Betelgeuse still guards many of its secrets well.

16 – ETA CARINAE

Discovering the fate of monster stars

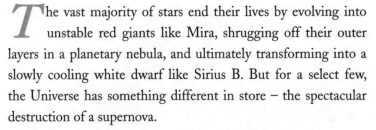

The vast majority of stars end their lives by evolving into unstable red giants like Mira, shrugging off their outer layers in a planetary nebula, and ultimately transforming into a slowly cooling white dwarf like Sirius B. But for a select few, the Universe has something different in store – the spectacular destruction of a supernova.

The key diagnostic for predicting this violent end is a type of morbid stellar obesity – if a star reaches the main sequence with a mass of more than eight times that of the Sun, then it's likely to go out with a bang unless other factors intervene. The red supergiant Betelgeuse, packing twelve Suns' worth of matter into its bloated frame, is doomed to this fate (probably in the next million years or so) but it's not quite there yet. Stargazers south of about 30°N, however, can feast their eyes on a star that's already showing signs of a fizzing fuse, and could blow at more or less any time.

At the time of writing, Eta Carinae is an indistinct star of middling brightness in the constellation of Carina, the Keel. Carina is a surviving fragment of what was once the largest constellation in the sky, Argo Navis. This was the ship that carried the hero Jason (Greek Tony Stark) with an ancient Greek super-team on the epic quest to <checks notes> find a yellowing sheepskin*.

The idea of a ship constellation was probably imported to Greece from ancient Egypt around the turn of the first millennium BCE (for Egyptians, this area of the sky was the Boat of Osiris, the multipurpose green-skinned god whose remit included agriculture and the afterlife). When the makers of the first modern star catalogues looked south in the early 1600s, however, Argo began to look a bit unseaworthy – packed with

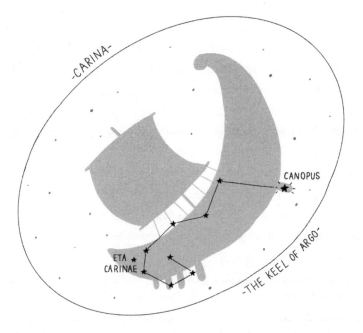

_____

* Okay, so the Golden Fleece was guarded by ferocious brass-hoofed bulls and an insomniac dragon, but it's still not exactly taking down Thanos, is it?

well over a hundred naked-eye stars, there were far too many for the Greek lettering system, devised by Johann Bayer, to cope with. In 1763, Nicolas-Louis de Lacaille, the incorrigible inventor of small boring constellations around the south celestial pole, catalogued Argo's stars by their position in the ship's sail, poop deck (don't laugh – it's the raised deck at the back), or keel. In the 1840s, the influential polymath John Herschel suggested a complete separation of Argo along these lines, giving rise to the modern constellations of Vela (the Sail), Puppis (the Poop Deck) and the one we're interested in, Carina*.

As the southernmost part of Argo, Carina is circumpolar for most of the southern hemisphere, rising highest in the mid-evening sky around April when the entire enormous ship hangs directly overhead. Skywatchers in tropical northern latitudes can see it highest in their southern sky around the same time – it's visible for several months to either side, but for just how long really depends on how far south you are.

At the time of writing, Eta Carinae appears as a magnitude 4-ish star – nothing that sounds too spectacular in its own right, but it's easy to spot because it lies at the heart of a bright star-forming nebula even more impressive than Messier 42, the home of Orion's famous Trapezium. Catalogued as NGC 3372[†], the Carina Nebula is even brighter than the Orion Nebula, and covers about four times the celestial real estate.

---

* Nevertheless, some astronomers continued to treat Argo as a unified whole until the official modern list of 88 constellations was drawn up in the 1930s.

† Charles Messier's Eurocentric catalogue of fuzzy celestial objects topped out at just over a hundred entries in 1784, but a century later Danish-born John Louis Emil Dreyer produced a massively expanded New General Catalogue (NGC) containing thousands of nebulae, star clusters and galaxies from all parts of the sky.

Through binoculars or a small telescope, the nebula is packed with detail – at a distance of about 7,500 light years from Earth it's almost six times further away than M42, which gives you an idea of how truly impressive it must be to stand out over such a great distance. Eta sits in a bright region of glowing gas hemmed in by two broad canyons of silhouetted dust that form a dark shallow V-shape across the face of the nebula (just above centre in the picture). Nearby, sits a smaller dark cloud called the Keyhole Nebula, while Eta itself shines out from a curious double-lobed cloud called the Homunculus Nebula (which we'll be coming back to later).

Eta makes it onto our tour of stars because, while it might be inconspicuous today, that wasn't always the case. The star was first recorded in 1677 by comet guru Edmond Halley, during a

sojourn on the remote south Atlantic island of Saint Helena*. Halley wondered why Ptolemy of Alexandria had skipped over it in his *Almagest* catalogue and toyed with the idea that it might have changed brightness since ancient times. This idea was borne out from the 1820s when Eta began to show an obvious increase in brightness[1].

John Herschel kept a close eye on Eta during an observing stint at South Africa's Cape of Good Hope, and recorded it brightening to around magnitude 0 by January 1838 (outshining Rigel in Orion, and nearby Alpha Centauri, before it began to fade slightly). After Herschel returned to England, he received reports from Thomas Maclear, Her Majesty's Astronomer at the Cape, chronicling a further outburst in 1843 that saw Eta briefly become the second brightest star in the entire sky at around magnitude -1.0[†]. After this, the star began to slowly fade, but remained at first magnitude for more than a decade before its brightness dropped off the proverbial cliff over a couple of decades from 1857 onwards. It faded below naked-eye visibility before a brief resurgence in 1887, and then a fairly steady period around magnitude six or seven that lasted for much of the twentieth century. Clearly, there was something unusual about Eta Carinae – while plenty of novae flared and faded, Eta's pattern of behaviour was completely different.

The complex surrounding nebulosity only added to the confusion. Throughout the late nineteenth century, the twin-lobed

---

*   Lately seized by the British East India Company after an unseemly squabble with their Dutch equivalent, and probably most famous today as a uniquely remote prison for Napoleon and other enemies of the rising nineteenth-century British Empire.

†   Overtaking magnitude -0.7 Canopus, which coincidentally lies at the opposite, western end of Carina.

shape now known as the Homunculus emerged and developed rapidly. At first, its apparently new appearance could be put down to improvements in observing tech – either nobody had been paying attention before, or they simply hadn't been able to see it. By the early twentieth century, many astronomers had concluded that what they were actually seeing was a double-star system.

But as it grew and brightened (helping to balance out the fading of the central star itself), it became clear that the new nebula was physically changing over time. In 1950, Argentinian astrophysicist Enrique Gaviola (a former pupil of Einstein, Bohr and other greats of early twentieth century physics) estimated the nebula's rate of growth by comparing photographs taken through a specially designed camera. He soon followed this up with spectroscopic measurements that measured the Doppler red and blue shift of lines in the nebula's spectrum, showing that expansion is driving the bubbles of gas that make up each lobe outwards, at speeds of hundreds of kilometres per second[*].[2]

This rate of expansion provides a neat way of estimating the distance to the Homunculus and, by extension, the entire Carina Nebula. By comparing the Doppler measurements of expansion with the rate at which the nebula's angular size in the sky changes (about 5 seconds of arc per century), you can figure out that it must be about 7,500 light years away.

In order to separate light from the star Eta itself from that of the surrounding Homunculus, astronomers have to use a trick called slit spectroscopy, which ensures that only a sample of light from the narrow area they're interested in is recorded onto a plate or sensor. When spread out into a spectrum of different wavelengths,

---

[*] Gaviola also coined the name Homunculus, suggesting the nebula had the shape of a crudely formed human being. Your mileage may vary.

Eta's light output is unlike almost anything else known, with a rainbow-like background "continuum" overlaid with bands of incandescent colour at specific wavelengths, and little or no sign of the dark absorption lines typically seen in a star's spectrum. Experts interpret this as a sign that light appearing to come from the central star has actually been absorbed and re-emitted by surrounding gas and dust – an energy recycling scheme that makes the spectrum very tricky to interpret.

This means that we must find other clues to the nature of the central star, and here we've been lucky. Having a good distance estimate to Eta shows that at the height of the 1843 eruption, the star must have been pumping out millions of times more light than the Sun – the kind of energy output that we'd normally expect from a supergiant like Betelgeuse, or a nova eruption. As astrophysicists got to grips with the complexities of stellar evolution in the twentieth century, they concluded that Eta was one of the most massive stars known – a rare type of supergiant called a Luminous Blue Variable (LBV).

You may recall from our look inside the Sun that when Arthur Eddington worked out the rules of stellar structure in the 1920s, he showed that any layer inside it would be kept in balance between the outward pressure of radiation from the core and the inward pull of gravity. Well there's an interesting corollary to this: increasing the mass of a stellar heavyweight (the kind of stars that shine by the super-efficient CNO fusion cycle, which we saw starting to take effect in Sirius) has a disproportionate effect on its energy output. Pile on a little more mass and, while the inward pull of gravity gets a wee bit stronger, the outward rush of radiation soars. This ultimately means that there's an upper mass limit to stars: a point above which a star will blow itself apart.

In extreme cases, that limit is absolute – try to create a single star with much more than a couple of hundred solar masses of material and you won't get far*. But when we're dealing with tens of solar masses, there's a grey area, where the most massive viable stars set off at the beginning of their brief but dazzling careers (limited to just a few million years by the availability of hydrogen in their core), only to have their plans altered along the way. These so-called Wolf-Rayet or WR stars† shed large amounts of mass (several Suns' worth in the course of their lifetime) through intense stellar winds that steadily strip away their outer layers and expose material closer and closer to the core. Their spectra become overlaid by intense emission lines – evidence of exposed surfaces that are already rich in helium, nitrogen and other products of CNO fusion.

Once a WR star runs out of fuel in its core and begins the "shell burning" phase (where most stars evolve into red giants or supergiants), the effects of this radical weight loss really begin to show. With most of the lightweight outer hydrogen envelope already gone, the star's rising luminosity is not accompanied by the usual dramatic increase in size and accompanying reddening: instead it may swell much less, becoming a luminous supergiant while its surface turns a hot yellow or white, or even remains blue in colour, potentially passing through a luminous blue variable phase.

Estimates put Eta's current surface temperature at around 40,000°C, indicating that it must pump out most of its energy

---

* The heaviest star we know of, the unpromisingly named R136a1, is a resident in the nearby Large Magellanic Cloud galaxy galaxy with a mass of about 265 Suns.

† Identified from their unusual spectra by Charles Wolf and Georges Rayet at the Paris Observatory in the 1860s.

as intense blue and violet light, along with copious amounts of invisible ultraviolet. But despite their name, LBVs can be changeable not only in brightness but also in colour. Although they spend most of their time in this hot blue state, they periodically go through a phase of expansion, in a cycle that varies from a few years to a couple of decades. This cools the surface and shifts more of the energy output into visible light, before the star shrinks back, heats up, and shifts into the ultraviolet once again.

If that mechanism's ringing a bell, it's probably because we saw something very similar happening in Mira and other red-giant variables (albeit at the cooler, infrared end of the spectrum). Astronomers think the driving force behind most LBVs is also similar to that at work inside Mira, with changes to the transparency of an internal layer triggering a cycle of expansion and contraction.

Despite this, the huge outburst staged by Eta Carinae in the 1830s was on an entirely different scale and likely due to very different forces at work – a sign of a star entering the final countdown to destruction.

We can't wind the clock back and see exactly what state Eta was in on the eve of its great eruption, but amazingly we can do the next best thing, and analyse the brilliant light from the explosion itself. Since the outburst predated the birth of spectroscopy by a couple of decades, the secret details hidden in this light might seem to be lost forever, but that's reckoning without the sheer ingenuity of some astronomers. In 2012, a team using the 4-metre Bianco telescope at Cerro Tololo Inter-American Observatory in northern Chile successfully captured the spectrum of light from the explosion, almost 170 years after it was first seen on Earth[3]. They did it by tracking down "light echoes" from the outburst – rays that ricocheted into their waiting 'scopes after setting off in a completely different direction and bouncing off a wall of gas elsewhere in the vast Carina Nebula. Untangling the spectrum of the starlight from that of the material reflecting it is a tricky task, but it reveals information that would otherwise be lost forever, showing for instance that Eta's outburst probably came when the star was in its cool yellow phase.

An LBV for northern skywatchers

While Eta Carinae is off the menu for observers north of the tropics, northern skies do offer a consolation in the form of P Cygni – an LBV that went through a similar outburst to Eta's in the early seveneenth century.

Embedded in the rich star clouds of the northern Milky Way, P Cygni is a fairly easy spot – look for the familiar cross-shape of Cygnus, the Swan. The intersection where the Swan's outstretched wings meet its body is marked by Sadr, a yellow-white star of magnitude 2.2, whose name comes from the Arabic word for "chest".

Follow a line south from Sadr, diverging a little to the east of the swan's official "neck", and within about five moon-widths you should fall over P Cygni. As a bluish star of about magnitude 4.8, it's easily visible under a fairly dark sky, but if you're struggling with light pollution, binoculars may help.

P Cygni made itself obvious when it became bright

enough to rival Sadr in 1600. It took several years to fade back, then made brief encores in 1655 and 1665, before settling to around its current brightness in the early eighteenth century. At about 5,500 light years from Earth, it may not benefit from Eta's spectacular surroundings, but it's worth keeping an eye on.

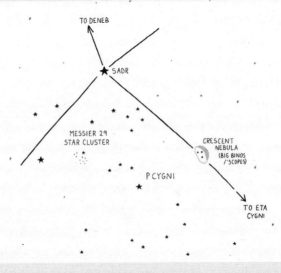

We saw in the case of Mira how a star with an exhausted core can produce not one, but two shells of fusion around it – the outermost burning hydrogen to produce helium, and the second transforming that helium into heavier elements such as carbon and oxygen. For most stars, this marks a last hurrah – a final blaze of nuclear energy before the outer layers dissipate and a burnt-out white dwarf is all that remains. Stars with more than eight times solar mass, however, can take things further – as the burnt-out, carbon-rich core collapses slowly inwards for a second time under its own weight, pressures and temperatures become so intense that the elements within the core begin to fuse in turn. The details of this process were first outlined by Yorkshire-born astronomer Fred Hoyle in 1954*, when he described how carbon and oxygen nuclei fuse with left-over helium atoms in various different combinations to build increasingly heavy elements such as neon and magnesium, silicon and sulfur[4]. These final stages of fusion ultimately generate a whole series of onion-skin fusion shells around the core, each moving outwards in slow but voracious pursuit of the products generated by its outer, lighter neighbour.

This process can only continue for so long, however – with each generation of new elements, there's less fuel to go around, and fusion itself releases less energy than before†. The supergiant becomes a stellar high-wire act, its multiple outer shells supported from beneath by increasingly feeble fusion taking place in the core.

Each phase of fusion is punctuated by a dormant period in which the star's core shrinks under the weight from above, until the engine sputters into life once again to produce a little more energy. Eventually, however – once the final stage of silicon fusion has built up a small core of iron and nickel – something breaks.

---

* The famously contrarian Hoyle later became better known for his sci-fi novels (most notably 1962's *A for Andromeda*, and for his championing of causes such as an eternal "Steady State" Universe (eventually disproved), his own model of gravity (a non-starter) and the theory of panspermia – abundant life in the Universe distributed between larger worlds by comets (the jury is still out).

† The reason for this has to do with the "nuclear binding energy" that holds the atomic nucleus together: fusion relies on the tendency in lightweight elements for binding energy to decrease with mass, so that fusing two atomic nuclei to form a third, heavier one leaves a small amount of spare energy to help keep a star shining.

As the star attempts to fuse iron, it comes up for the first time against an element that *absorbs* energy during fusion, rather than releasing it. The results are spectacular – as the outward pressure of radiation from the core cuts out, the vast weight of the overlying star abruptly comes crashing down onto it, before rebounding outwards in the vast stellar explosion known as a supernova. While stars with masses of 8 to 40 Suns will have swollen into red supergiants in the run-up to this moment, those with even greater masses have their expansion constrained by a combination of gravity and mass loss, and approach their violent end as LBVs .

We'll look in more detail at supernovae – and what they leave behind – in the next chapter, but for now it's enough to say that Eta is hurtling towards just such a fate, and becoming increasingly unstable as it approaches it. Estimates put the total mass lost during the years of eruption that peaked in 1843 at a staggering 20 Suns (most of it going to form the expanding Homunculus Nebula), which raises the obvious question of just how heavy this star really is.

Until the 1990s, the general assumption was that Eta was a single monstrous star with a mass of at least 100 Suns. In the late twentieth century, however, as the gas and dust thrown out in 1843 gradually thinned, Eta began to steadily brighten again. And as astronomers plotted the star's slowly rising "light curve" using the latest instruments, they began to suspect a pattern beneath at least some its unpredictable variations – a more complex version of the fade-outs caused by periodic eclipses in binary stars like Algol. In 1996, Brazilian researcher Augusto Damineli analysed related changes to Eta's spectrum and suggested that Eta might, after all, be a double star. His theory has since been backed up by an array of studies at different wavelengths, although the two stars are far too close to separate with even the most powerful telescope[5].

Estimates of the masses for the two monster stars vary wildly, but a conservative assumption would still put the primary at 100+ Suns, with the smaller companion weighing in at around 45 Suns. Powerful stellar winds blowing out from the surface of each star collide in between them, creating a million-degree-plus "hotspot" that emits intense and dangerous X-rays. Once in each orbit, the hotspot is eclipsed from Earth, causing the X-ray source to briefly blink off and on.

While the fact that Eta is a binary system adds yet further complexity to this puzzling star, and makes its behaviour even trickier to predict, its long-term fate is sealed. Both stars, despite being stellar toddlers still firmly embedded in the stellar nursery that created them, are on the fast track towards an early death. The unstable primary star, in particular, could potentially blow at any time in the next 100,000 years or so, producing an explosion that will at the very least rival Venus, and could (depending on the star's precise mass and geometry) be up to 20 times brighter in Earth's skies. Eta's outburst of 1843, now classed as a "supernova impostor event", was just a preview of forthcoming attractions.

17 – The Crab Pulsar

A famous supernova, and what it left behind

E arly on the morning of July 4, 1054, Yang Weide, court astronomer to Emperor Renzong of China's Northern Song dynasty, noticed a change in the morning sky. Sitting directly between the waning crescent moon and the glow of an approaching sunrise blazed a new star – one so bright that it outshone even the brilliant Venus. For 23 days, the star remained visible even in daylight, and after that it faded only slowly, remaining on show in the night sky for some 20 months.

Today's astronomers give this cosmic spectacle the somewhat dull designation SN1054, but to its friends, it's the Crab Supernova. As one of the brightest stellar explosions in recorded history, it's conservatively estimated to have reached a peak magnitude of -6, and it left behind an expanding cloud of superheated, shredded gas that we can still see today, with a secret buried in its heart.

The trail of SN1054 leads us back to the mighty constellation of Taurus for one last visit. This most recognizable and ancient of constellations was the Bull of Heaven in ancient Mesopotamia, sent by the goddess Ishtar to confront the hero Gilgamesh who had spurned her advances. In the Classical world, meanwhile, it was generally seen as Zeus, up to his usual zoomorphic tricks on his way to abduct the Phoenician princess Europa.

The new star appeared just to the northwest of Zeta Tauri, an obvious star of magnitude 3.0 that marks the tip of the bull's southern horn. Just as it did a thousand years ago, Taurus first becomes visible in morning skies as the Sun, making its annual eastward loop around the sky, moves into the neighbouring zodiac constellation of Cancer in late June. By October it's rising in the middle of the evening and it remains obvious part of the night sky until it disappears into the sunset in April.

The scene of the incident is today cordoned off by the famous Crab Nebula. You'll need good binoculars and a clear dark sky (or a small telescope) to see it, but once you've zeroed in on Zeta Tauri, it should be pretty obvious – a blurry patch of light

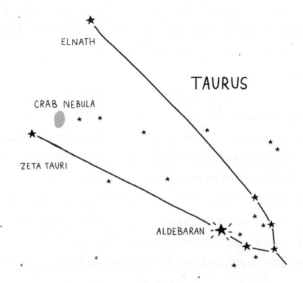

about a quarter the diameter of the Full Moon, and a couple of moon-widths away from the star. A small telescope and good magnification may show you individual filaments of glowing gas around its edges.

The Crab was first discovered in 1731 by London-based doctor and part-time astronomer John Bevis, but it only became famous a generation later when French stargazer Charles Messier stumbled across it while searching in 1758 for the reappearance of Halley's Comet. Gaining top billing on his famous catalogue of irritatingly comet-like objects as "Messier 1", the nebula soon became a popular target for stargazers – and the frequent subject for debate about whether it was made of gas or a vast cloud of unresolved stars.

The Crab's familiar name was coined a few decades later by William Parsons, the 3rd Earl of Rosse. This wealthy member of the Anglo-Irish nobility eyeballed the nebula through an enormous 72-inch reflecting telescope know as the Leviathan of Parsonstown – a monster that rose from the grounds of his Birr Castle home in the late 1840s, even as the infamous potato famine ravaged the surrounding lands. So unwieldy that it could only be tilted up and down on a single axis (and was therefore limited to observing whatever was passing through a particular strip of the sky at a given moment), the Leviathan was nevertheless one of the premier scientific instruments of its day, and allowed Parsons, teetering on a precarious high-level observing platform, to see the nebula's filament-like structure for the first time[1].

While the first photographs of Messier 1, taken in the 1890s, did away with the lingering idea that it might be a cloud of stars, the Crab's true nature didn't become clear until 1913, when Arizona-based Vesto Slipher (of whom we'll be hearing a lot more later) succeeded in capturing its elusive spectrum. This turned out to be a real headscratcher – a continuous faint background with a number of bright emission lines, each of which turned out, on closer inspection, to have two peaks of intensity offset to either side of their expected position. It took some time for Slipher to realise he was seeing emission lines from both sides of a vast expanding cloud of energized gas, pushed in different directions by the Doppler effect: the Crab Nebula, it seemed, was growing rapidly.

This expansion was confirmed by other means in 1921 when Slipher's Lowell Observatory colleague Carl Lampland published the results of a laborious trawl through the photographic archives. Snapshots taken over almost a decade showed beyond doubt that the nebula was changing over time[2].

A link between the nebula and the explosion of 1054 was first suggested by Edwin Hubble in 1928. We'll be seeing just why he's a big deal in the history of astronomy when we reach Andromeda in a later chapter, but in the meantime, his observation about the Crab is a rather brilliant throwaway. Writing an introductory article on novae, Hubble noted that – based on the measured rate of the nebula's expansion and its current angular size – you could rewind through time and see that it originated about 900 years ago. Since the nova of 1054 was the only known outburst in Taurus at around that time, it seemed reasonable to link the two[*3]. Hubble never returned to the topic, but within a couple of decades other astronomers had sealed the deal.

★ ★ ★

* Credit should also go to Sweden's Knut Lundmark, who in the early 1920s did the hard work of sifting through ancient Chinese records of "guest stars" and compiling the first catalogue of possible historic novae.

Hubble, you may have noticed, referred to the explosion as a nova rather than a supernova – in fact until the 1930s, astronomers didn't really make the distinction. The occasional stars that flared into view and gradually faded away were treated as a single type of object, and since no one knew much about them anyway, it hardly mattered.

The seeds of change, however, had been planted in 1885 when a bright star flared up in what was then called the Andromeda Nebula. S Andromedae, as the eruption was known, barely reached naked eye visibility at its peak, but it took on a new significance after breakthroughs in the measurement of cosmic distances (spearheaded, coincidentally, by Hubble) during the late 1920s. If you don't want to know the results for Imposter 2, look away now…

When the Andromeda Nebula turned out to be an independent galaxy more than 2 million light years from Earth, S Andromedae was catapulted into a different class from a common or garden nova. In 1934, German Walter Baade and Swiss Fritz Zwicky, both working at California's Mount Wilson Observatory, figured out that at its peak, the explosion was at least a million times brighter than the Sun – and coined the term "super-nova" to describe it[4].

Not many scientific papers really deserve the adjective "seminal", but Baade & Zwicky (1934) is probably one that does. In the space of a mere six pages they set out the case for supernovae as a distinct class of explosion, and produced ballpark figures for telling the two apart (estimating that novae peak at a maximum of about 20,000 times the brightness of the Sun, while supernovae are about 50 times brighter than that). They went on to identify S Andromedae and the famous nova of 1572 (see RS

Ophiuchi) as members of the supernova class, and pointed out that unlike normal novae, the stars involved in supernovae could not be detected in the aftermath of the explosion. Through a combination of guesstimation and mathematical derring-do, they even put a rough figure on the vast amount of energy liberated during a supernova explosion.[*]

For a final flourish, the paper included a calculation for the most efficient means of energy generation possible – the direct conversion of mass into energy à la Einstein's $E=mc^2$. Even erring on the conservative side, this revealed that a supernova explosion was the equivalent of converting a solar-mass star *completely* into energy. The authors rightly concluded that a supernova was a rare transition of a high-mass star into an object with far smaller mass.

Baade and Zwicky were keenly aware that their dataset for supernovae was currently limited to just two items, and that they'd probably die of boredom if they just sat around waiting for another one to happen on our cosmic doorstep. So, together with their colleague Rudolph Minkowski (like Baade, another German-born immigrant to the US), they began a survey of remote galaxies, reasoning that if the Milky Way produces one supernova every few centuries[†], then keeping an eye on a few hundred galaxies might yield one a year.

The search was more successful than the three men had dared to hope, and by 1941 they had more than a dozen supernovae under their belts. This allowed Minkowski and Zwicky to

---

[*] The figure the came up with involves an awful lot of zeroes (48 of them, to be precise) and a none-too-familiar unit, the erg. In a more easily understood shorthand, they estimated that a supernova releases as much energy in a few months as the Sun does in 10 million years.

[†] More recent calculation suggests they should average once every 50 years in our galaxy, so considering that the last galactic supernova was in 1604, we are *way* overdue...

distinguish between several distinct types of explosion according to features of their spectrum and the light curve describing their rise and fall in brightness*. Baade, meanwhile, had also been scouring historical records in search of possible neglected supernovae, and in 1942 this led him to confirm that the new star of 1054 was not just a nova but a supernova, with the Crab Nebula as its rapidly expanding remnant[5].

★ ★ ★

As we saw on our visit to Eta Carinae, supernovae come about when a monster star with more than eight times solar mass reaches the end of its life. The previous phase has been concerned with the development of a complex onion-skin structure of thin shells deep beneath a surrounding envelope of hydrogen gas. Each shell is a separate fusion factory, but when the core fills up with unusable iron and nickel, the main stellar power source is abruptly cut off. Suddenly robbed of the radiation pressure that has previously held them up, the star's delicately balanced layers collapse faster than a top-heavy motorcycle display team, plunging towards the core at speeds of up to 70,000 kilometres per second.

This initial implosion doesn't last long, however – as the innermost shells strike the collapsing iron-rich core, they rebound to produce a cataclysmic shockwave that tears its way out through the star. Anything that gets in the way of this shockwave is compressed and heated to immense temperatures, triggering a vast wave of nuclear fusion at temperatures far above those found

---

* The most important distinction for our purposes is between "Type Ia" supernovae and all the rest. While most supernovae are exploding stars of the type covered in this chapter, Type Ia are something else entirely (see Supernova 1994D for more details)

in the cores of even the biggest supergiants. With so much excess energy around, fusion can even create nuclei heavier than iron – elements that are normally "off limits" due to their tendency to absorb, rather than release, energy during their formation. As a result, the star burns through several Sun's worth of material in a matter of days, before scattering its shattered wreckage across the neighbourhood like a cosmic fly tipper.

That's the short version, at least – the reality is a wee bit more complicated, since computer models suggest that most of the energy from the initial shockwave is expended deep beneath the star's surface as its passage breaks apart all those heavy-element nuclei that have been so painstakingly created over the past million or so years. *Something* then happens to re-energize the shockwave so that it can continue out through the star and create a whole new smorgasbord of heavy elements from scratch. Just what that something is, is related to what's going on in the core.

* * *

You may remember from Sirius B that a white dwarf is the final stage in the life of a star with roughly the Sun's mass. This burnt-out, Earth-sized stellar core has enormous density (imagine an elephant squeezed into a matchbox), and is supported against further collapse by a strange kind of pressure that works at extremely short range to repel individual electrons from each other. Well, in 1931, Subrahmanyan Chandrasekhar, a young Indian astrophysicist steeped in the recent developments of quantum physics, theorised there must be an upper limit to the mass of a white dwarf, where this pressure would fail and the burnt-out star would undergo a complete collapse (which he thought would end in the formation of a black hole). Chandra (as he's generally known) calculated the cut-off point to be about 1.4 masses. Since a star needs a total mass of at least 8 Suns if it's to scale the later stages of fusion and produce a supernova, it's likely that the core of any exploding star will be above this "Chandrasekhar limit".

What Chandra couldn't have realised at the time (but Baade and Zwicky correctly guessed a couple of years later) is that there's a safety net waiting to catch the collapsing white dwarf before it crumples into a black hole. This is called neutron degeneracy (jargon from quantum physics, not a moral judgment).

Finally confirmed in 1933, neutrons were the last of the three main types of subatomic particle to be discovered. If protons are the important ones whose numbers decide which element a particular atom belongs to, and electrons are the useful ones that whizz around the outside and allow the atom to interact with others, then neutrons are the makeweights, chumming up with protons in the atomic nucleus and usually not doing very much except adding mass. However neutrons have a secret — in certain circumstances they can change into protons and electrons, and in *extreme* circumstances (for example, inside a collapsing stellar core), protons and electrons can turn into *them*.

Thus, as the supernova's core shrinks past the size of the Earth down to the size of the Moon and beyond, electrons and protons are rammed into each other with incredible force, forming neutrons and releasing near-intangible particles called neutrinos*.

---

* Neutrinos are also emitted as a way for the neutron star to rapidly shed the vast amounts of excess heat generated during its collapse, and this is thought to be what re-energizes the supernova shockwave. Almost massless, effectively unstoppable and moving very close to the speed of light, they escape from the explosion at the moment of its conception. If they can be detected, they can provide a cosmic early warning system that a supernova is afoot.

Eventually, the core is nothing but neutrons – and just as we saw the Pauli exclusion principle appear at the last moment to save Sirius B from collapse, so it reveals itself again here, stepping in to create resistance between neutrons as they cram tightly together.

The resulting neutron star is a city-sized object that packs the mass of a fully laden supertanker into each pinhead of material, with a surface temperature of a million degrees plus. Its surface screams out radiation across a broad range of the electromagnetic spectrum from X-rays to visible light and beyond. But as we saw with white dwarfs, temperature alone isn't a guarantee of visibility in the wider Universe – a star also needs a certain size if it's to be seen. The idea of spotting objects a few kilometres across over many light years of space appeared impossible at first, and for a long time it seemed that neutron stars were doomed to remain in the realms of speculation, forever beyond the limits of even the most powerful telescopes.

That all changed in November 1967, when a young PhD student called Jocelyn Bell stumbled across something strange in a field outside Cambridge. The field in question had been filled with hundreds of radio antennae comprising the Interplanetary Scintillation Array – an advanced radio telescope designed to pin down the position of unpredictable radio sources called quasars (more on these when we get to one of the most famous, 3C 273). However, Bell, poring over the printouts of data gathered the previous August, discovered something completely different – a signal pulsing from the sky with metronomic regularity.

Initially nicknamed LGM-1 (short for Little Green Man), the strange signal sat in the constellation of Vulpecula, the Fox, and pulsed every 1.337 seconds – so quickly that it could not be caused by any known star. Bell fought to convince her supervisor Antony Hewish that the signals were natural and not caused by artificial interference, and eventually Hewish put his reservations to one side. The first rapidly pulsating radio source (later known as a pulsar) was announced under the fairly dull monicker of CP 1919[6].

By sheer coincidence, this happy event came mere weeks after Italian researcher Franco Pacini had published a model of how neutron stars could create just this sort of phenomenon[7]. Thanks to the conservation of angular momentum (a physical law that we previously encountered shaping the birth of stars like T Tauri), compressing a star's core into a volume the size of Birmingham causes it to spin up like a hyperactive washing machine. At the same time, the star's original magnetic field is compressed to a fraction of its normal size with a huge accompanying boost in intensity. The two effects together, Pacini showed, would funnel any radiation from the star's surface and surroundings into a pair of powerful, tightly aligned beams shooting into space from each magnetic pole. Since the beams were unlikely to align precisely with the axis of rotation, they would sweep around the sky like a cosmic lighthouse, producing rapid flashes for anyone who got in their way.

But however neat the match between theory and observation, science tends to require more evidence for its breakthroughs than a single set of radio signals from a lone object, even one as iconic as CP 1919*. The discovery of more pulsars with similar characteristics certainly helped, but it was the Crab Nebula that would provide the clinching evidence. Barely a year after Bell's initial discovery, astronomers using the Green Bank Radio

---

* Gracing the sleeve of Joy Division's 1979 debut *Unknown Pleasures*, the pulsar's radio trace has probably appeared on more T-shirts than that Nirvana baby.

Telescope in West Virginia and the vast dish antenna at Arecibo in Puerto Rico detected a pulsar signal from the Crab, repeating an incredible 30 times per second. The discovery provided the missing link between pulsars and supernovae, and given the recent date of the Crab explosion, suggested that pulsars start off fast and gradually lose speed and strength over time[*].

Since the discovery of the Crab Pulsar, it's been the subject of unrelenting study. In 1969, astronomers confirmed that it had been hiding in plain sight all along, when they found that one of two faint stars coinciding with the centre of the nebula was flashing every 33 milliseconds. Ironically, this flashing had been reported by sharp-eyed observers several times over the years before, but dismissed by know-it-all professionals. More recently, spectacular snaps from Hubble and other satellites have shown the hidden structure carved out by the pulsar from within the Crab Nebula, with jets of matter streaming from either end of an out-of-control spinning top to energize their surroundings.

There's a famous aphorism (often attributed to Geoffrey Burbidge, who worked with Fred Hoyle and others to explain the origin of heavy elements within supernovae) that astronomy can be divided into the study of the Crab Nebula and the study of everything else. Given how much the Crab and its pulsar have revealed about the way stars die, it's easy to see why it's stuck.

---

[*] Hewish and radio astronomy pioneer Martin Ryle got the Nobel Prize for the discovery while Bell, somewhat infamously, was overlooked.

18 – CYGNUS X-1

*Searching for a black hole in
a dark Universe*

✳

*A*nd now we arrive at one of the strangest objects in the cosmos – a thing that, by definition, we cannot see directly, but which nevertheless loiters around the sidelines of astrophysics, occasionally knocking things over and making trouble like an unruly cosmic poltergeist. In other words, a black hole.

A black hole is, put simply, an object with gravity so strong that nothing, not even light travelling at 300,000 kilometres per second, can escape its clutches. Anything that comes too close and moves too slowly is inevitably pulled to its doom, but otherwise, contrary to what some science-fiction movies would have you believe, it just sits there and glowers invisibly at the surrounding Universe.

The particular black hole we're interested in, known as Cygnus X-1, has a special place in history – as the first object of its kind to be discovered, it catapulted a theoretical curio into reality. Looking for an object you can't see might sound like a quest that Lewis Carroll would set for Alice, but fortunately X-1 isn't alone in space – it has a companion star whose behaviour gives away the black hole's presence.

It doesn't take a genius to figure out that X-1 lies within the bounds of the constellation of Cygnus, the Swan. Flying highest overhead in northern skies on August and September evenings, Cygnus stretches its neck down a bright stretch of the Milky Way towards the southern horizon. For southern-hemisphere skywatchers, meanwhile, it appears to soar above the northern horizon in the same months.

The star we're looking for, cursed with the unglamorous designation of V1357 Cygni, lies more or less halfway along the Swan's neck, though you'll probably need a small telescope to spot it. Look midway between the bright star Sadr on the swan's chest, and Albireo, the beautiful double star that marks its beak, and you should find Eta Cygni, unassuming at magnitude 3.9.

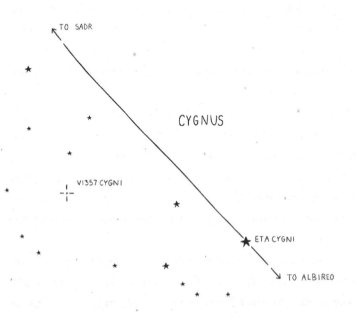

V1357 lies about a moon's-width to its northeast. You can follow the sketch to track it down but it's nothing very spectacular to look at – just a bluish star of average magnitude (about 8.9) that puts it right on the cusp of visibility for a pair of good binoculars.

That boring appearance, however, belies a star that's rather impressive in its own right. With a spectral type of O9, it belongs at the hot (and therefore highly luminous) end of the main sequence band of stars. For reasons we've explored in previous chapters, this means it's likely to have a high mass (estimated at about 18 Suns), and a relatively short lifespan of perhaps a few million years. The star's varying brightness, meanwhile, is due to a mechanism that we haven't encountered before – it's one of a fairly rare class known as ellipsoidal variables. These are stars with a distinct bulge around the midriff that display different amounts of their surface to us at different times. In most cases, this type of variability is created when a star spins so rapidly that its equatorial regions are attempting to fly away into space, but as we'll see in this case, the distortion is largely due to the influence of the nearby black hole.

Finally, it's worth noting that V1357's proper motion through space matches pretty well to a group of stars called the Cygnus OB3 association – a loose stellar grouping with a common origin, similar to the Ursa Major Moving Group of which Mizar is a member. Cygnus OB3 is thought to have formed about 5 or 6 million years ago, which will become relevant later in our story.

Astronomers in the early 1970s were drawn to this not-particularly-remarkable star by something that *is* remarkable – one of the strongest X-ray sources in the sky. Most of us are only familiar with X-rays from the domesticated form (usually encountered from the discomfort of a dentist's chair), but these magic rays that pass through flesh and bone with insulting ease are also a natural phenomenon. They're part of the same electromagnetic spectrum as normal light, infrared and ultraviolet, but have much shorter wavelengths and higher frequencies than any of them. This means they carry much more energy, and can only be produced by processes with a lot of *oomph*.

Usually, this oomph comes in the form of heat – we've already seen from stars like Eta Carinae that gas heated to tens of thousands of degrees will produce ultraviolet rays in preference to visible light, and if you can somehow push beyond that to a million degrees or so, you can start to generate X-rays in quantity*.

The fact that the Universe is chock full of material at these kinds of extreme temperatures only became clear in the 1960s, when X-ray detectors were taken to the edge of space aboard slender suborbital rockets called Aerobees. These early forays into space-based astronomy revealed that the sky was scattered with X-ray sources at a variety of different wavelengths and energies (many, for instance, are produced by superhot gas in the Sun's outer atmosphere). Fortunately for those of us who aren't keen on constant irradiation, nearly all of these space rays are blocked by Earth's atmosphere.

Cygnus X-1, the third brightest X-ray source in the sky, was discovered on one of these scientific thrill rides in 1964, but the detector system (which relied on a Geiger counter peering out

---

* Laboratory (and surgery) X-ray sources, by the way, create small numbers of the high-energy collisions needed to produce X-rays by accelerating charged electrons through electric fields with very high voltages.

of a window in the side of a slowly spinning rocket like a budget airline passenger craning their neck to spot an Alp) could only pin it down to a broad area of the sky[1]. By 1970, the launch of NASA's first dedicated X-ray satellite, Uhuru, had helped narrow the search area, but also deepened the mystery of the source itself.

Cygnus X-1 turned out to be highly variable, changing its intensity several times a second. Because its brightness jumped and dipped suddenly rather than varying smoothly, the X-ray source also had to be pretty small. Basic physics says that physical changes can spread through an object no faster than the speed of light, so sudden changes put a limit on an object's size. In Cygnus X-1's case, the X-rays had to be coming from a region less than 100,000 kilometres across.

Casting a wide net, ground-based astronomers soon got busy looking for a possible counterpart in visible light. At first, they drew a blank – nothing in this part of the sky particularly screamed for attention. But then in 1971, radio astronomers from Leiden University in the Netherlands and the US National Radio Astronomy Observatory, independently discovered that Cygnus X-1 was also producing radio waves. The source of these signals was a lot easier to pin down than the X-rays, and there, waiting to be discovered, was the star now known as V1357 Cygni.

It soon became clear, however, that V1357 could not be the source of the rays itself. This perfectly normal star might be a little on the big, bright and hot side, but it was utterly incapable

---

* Uhuru is Swahili for "freedom" – a nod to the satellite's Kenyan launch site. And yes, Trekkies, the same word inspired the naming of the USS *Enterprise*'s iconic comms officer, Lieutenant Uhura, a few years before.

of launching X-rays across hundreds of light years of space. So perhaps there was something else?

As is often the way, the solution occurred to two independent teams at once. In the autumn of 1971, Louise Webster and Paul Murdin, both of the Royal Greenwich Observatory*, and Tom Bolton at the University of Toronto both set out to measure the spectrum of the visible star, hoping to find periodic Doppler shifts in the starlight as it was pulled back and forth by an invisible X-ray source[2]. What they found surprised and excited them all – the Doppler shifts were not only present, they were strong. So strong, in fact, that simple orbital models of the system suggested the source had six or more times the mass of the Sun[3]. An otherwise invisible object with a mass like that could only be one thing.

* * *

The possibility of something like a black hole existing was suggested remarkably early in the story of astronomy – in 1783 to be precise. That year John Michell, a clergyman and philosopher with something of a talent for foresight, presented a paper to London's Royal Society outlining the basic idea of a "dark star" whose gravity was strong enough to hold onto its light[4]. Unfortunately, Michell's idea was lost to obscurity for almost two centuries, and so, when the idea resurfaced in 1915 as a possible consequence of Einstein's famous theory of general relativity, Michell's prior claim was overlooked.

---

* Established in 1675, the RGO had moved in the 1950s to Herstmonceux, Sussex, with its former Greenwich site renamed the Old Royal Observatory. It later moved again to Cambridge before being shuttered in 1998, whereupon the original site was renamed the Royal Observatory Greenwich, which isn't the same thing at all.

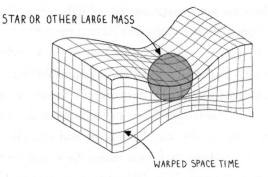

STAR OR OTHER LARGE MASS

WARPED SPACE TIME

So, a quick – and hopefully painless – word on general relativity. The basic idea is that what we think of as the three separate dimensions (directions) of space and one of time are in fact connected in a sort of flexible, four-dimensional mesh (which Einstein's former tutor Hermann Minkowski named "spacetime") . The mesh gets pinched and distorted around large masses, deflecting the paths of objects moving past them and creating the effects we experience as gravity. The whole set of relationships is described by a set of mathematical formulae called the field equations.

When German astronomer Karl Schwarzschild investigated these equations shortly after their debut, he found that nothing within them expressly prevented the existence of curious points called singularities: plug the right set of numbers into the equations, and they could produce a single point in space with infinite density*. Furthermore, what seemed at first to be a bug in the theory was actually a feature: the very properties of singularities that seemed like they would wreak havoc with the laws of physics *also* mean that they seal themselves

---

* Think of a singularity as relativity's version of breaking your pocket calculator by typing in "divide by zero".

away from the Universe, keeping out of mischief behind an impenetrable barrier from which not even light can escape[5]. Sound familiar?

Schwarzschild also demonstrated that if you compressed any massive object down below a certain size (now known as its Schwarzschild radius), you could create a singularity, but his ideas were largely seen as a mere curiosity of the mathematics – safe to ignore since no one could imagine a way in which such a singularity might form.

That all changed in 1931, when the young Subrahmanyan Chandrasekhar (who we met in our previous chapter on the Crab Pulsar) discovered a way of making the damned things. Chandra, whose uncle had been the first Indian winner of the Nobel Prize for Physics, was en route from his homeland to begin doctoral research at Trinity College, Cambridge when he had the revelation that singularities might arise naturally from the collapse of stellar cores above 1.4 solar masses. In these extreme examples of stardeath, powerful gravity would prevent the formation of a stable white dwarf, and so the core would continue its collapse to become a single superdense point. Chandra's idea soon brought him into contact, and conflict, with the greatest astrophysicist of the time.

Arthur Stanley Eddington had made his reputation by championing general relativity and leading a 1919 expedition that provided a stunning demonstration of the theory in action. In the 1920s, he had developed ideas about stellar structure that form the bedrock for much of the science in this book. And now in the 1930s, he was turning his attention to ways of combining relativity with the strange new science of quantum physics. You'd think Chandra's ideas should have been right up his street, and yet he not only rejected them – he did so with prejudice, using the weight of his reputation and a considerable rhetorical talent to make the youngster's theory seem ridiculous and unacceptable[*]. Quite why he took against black holes so vehemently is unclear: some have suggested it conflicted with ideas that he was developing, others that he had an almost philosophical objection to the idea of singularities and stars disappearing out of existence. Chandra himself believed that racism played a part, and ultimately quit the suffocating atmosphere of Cambridge for life in Chicago.

In the mid-1930s, the discovery that neutron stars (such as the Crab Pulsar) could form from the collapse of high-mass stellar cores appeared to offer room for compromise – perhaps black holes weren't necessary after all? But the arc of science (at least) is long and bends towards justice. By the end of the decade, the reality of singularities was back on the agenda thanks to work by Robert Oppenheimer (yes, the Manhattan Project guy) and Russian-Canadian physicist George Volkoff. Using work by mathematician Richard C. Tolman, they made the first estimate of the mass at which even neutron degeneracy pressure (again, see the Crab Pulsar) would give up in disgust – the so-called Tolman-Oppenheimer-Volkoff (or TOV limit)[6†].

---

[*] Ironically Eddington made his own important contribution to the physics of singularities before turning against them – he pointed out that because the speed of light is fixed, it cannot really be "slowed down" by the pull of gravity – instead, light escaping from near the boundary has its wavelengths stretched and is eventually red shifted into invisibility – a so-called "gravitational redshift" that has now been confirmed by observation of stars such as S2.

† About 2.3 Suns, if some recent measurements are reliable, suggesting there's a relatively narrow gap for neutron stars to fit into the picture, and any star with a mass above about 18 times solar is ultimately black hole bound.

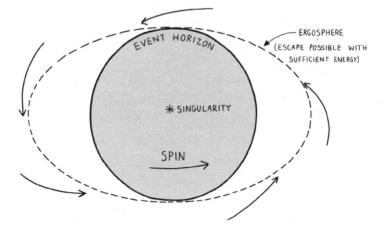

ANATOMY OF A BLACK HOLE

EVENT HORIZON

ERGOSPHERE
(ESCAPE POSSIBLE WITH
SUFFICIENT ENERGY)

✳ SINGULARITY

SPIN

It wasn't until the 1960s, however, that the golden age of black hole physics really kicked off. The opening act came in 1958, when David Finkelstein, working at the Stevens Institute of Technology in Hoboken, NJ, realised that the Schwarzschild radius not only described the point of no return in black hole formation, but also how it would appear forever after: even as the material inside dwindles to a singular point, a barrier called the event horizon (the surface where gravity becomes so intense that light is trapped) remains in place around it[*].

Whole books have been written about the mind-bending physics that began to pop up as mathematicians and astronomers looked more closely at the event horizon, but time is pressing and it's almost time to get back on our whistle-stop cosmic coach tour. For the moment, it's enough to say that by the early 1970s,

---

[*] To some extent, the event horizon lets physicists off the hook – with the singularity safely sealed off and no chance of its influence leaking out into the rest of the Universe, they can safely ignore whatever cosmic rulebreaking is going on in inside.

black holes had been revealed as far more complex objects than previously suspected*. The fact that they also remained entirely theoretical was becoming a little bit embarrassing, though.

The discovery of Cygnus X-1, therefore, seemed perfectly timed, even if some of the theoreticians remained cautious. Stephen Hawking, for instance, made an insurance bet against himself, wagering with his friend Kip Thorne that Cygnus X-1 would turn out *not* to be a black hole†.

The big question, of course, is what exactly was creating the X-rays? A black hole is, by definition, black, and while it might conceivably red-shift the light from orbiting objects into infrared and radio waves as they spiraled to their doom, the object itself has no real business producing high-energy X-radiation.

Fortunately, theoretical astrophysics had a model just waiting for situations such as this – the "accretion disc". First proposed by German nuclear scientist Carl Friedrich von Weizsäcker in the late 1940s, this hypothesis suggested that gas falling onto a rotating object (for instance when mass transfers between stars) does not simply accumulate directly on the surface – instead, as particles drifting in different directions jostle and collide with each other, they tend to even out and form a disc rather like those we saw involved in star formation[7] (see T Tauri). Once "accreted" onto the disc, particles of matter follow inward spiral

---

* They'd also found their name – no one knows who first came up with the term "black hole", but it had certainly entered the lingo by January 1964, when it pops up in a report by journo Ann Ewing on the American Association for the Advancement of Science's annual jamboree, in Cleveland Ohio.

† Hawking stood to win a four-year subscription to *Private Eye* as a consolation prize should Cygnus X-1 disappoint. When he happily conceded the bet almost two decades later, Mrs Thorne was apparently none too impressed at her husband's side of the wager – a year's subscription to *Penthouse*. The 1970s, folks.

tracks like a needle on rare vinyl until they eventually reach the surface of the attracting object.

As matter spirals down, it can heat up due to simple friction with its surroundings (and also, it's now known, through interactions with magnetic fields). When the extreme gravity around a stellar remnant is added to the mix, temperatures in the disc can reach millions of degrees, causing it to radiate with energies well into the range of X-rays.

Decades of further study have confirmed beyond doubt that Cygnus X-1 is surrounded by just such a disc – its radiation is a last superheated shriek from matter before it is pulled onto the black hole's event horizon. In 2011, astronomers using simultaneous observations from three separate astronomy satellites were able to probe the disc's properties in unprecedented detail, measuring the black hole's mass at an impressive 14.8 Suns, and confirming that it is spinning 800 times per second – a figure that allowed them to date its origin to around 6 million years ago[8]. This means that its progenitor star must have been been a true monster that lived and died in the earliest days of the Cygnus OB3 association.

Since the identification of Cygnus X-1, similar X-ray discs have been used to pinpoint scores of stellar-mass black holes in our galaxy and beyond, transforming Michell's dark stars from theory to undeniable reality. When we come to 3C 273 in our penultimate chapter, we'll be seeing an example of the same processes at work on an even more awesome scale.

19 – Eta Aquilae

*Cepheids – a measuring tape
for the cosmos*

✦

Many stars are variable, but some are more variable than others. Some follow the celestial clockwork of orbits and eclipses, others vary fairly predictably as their rotation brings dark starspots or upwellings of hot, bright material into view. Some shift their energy output dramatically from the visible to the invisible and back as they as they vary in size and surface temperature, while others belch out brilliant flares or clouds of obscuring gas and dust seemingly at random.

But one class of stars are the last word in variables – not only do these stars follow a predictable cycle, but the periods of individual stars of this type can reveal their other properties, offering a way to find the distance to objects far beyond the reach of traditional parallax measurements, and presenting a chance to put a scale on the entire Universe. These variables – an astronomical equivalent of Douglas Adams' improbably useful Babel fish, are known as Cepheids. As a class they're named after Delta Cephei, a

relatively bright star in the far northern constellation of Cepheus*
that varies between magnitudes 3.5 and 4.4 in a period of 5 days
8 hours and 53 minutes.

Delta Cephei's periodic wobbles were first spotted by John
Goodricke (discoverer of Algol), and announced to the world
on New Year's Day 1786[1]. However, Goodricke's friend and
collaborator Edward Pigott had in fact found a star with very
similar behaviour in the rich star fields of the constellation Aquila
the previous year[2]. Since this star is the true prototype of the
"Cepheids" and Aquila is a rather more interesting and widely
accessible constellation, in this chapter we'll focus on Pigott's
discovery – Eta Aquilae.

Aquila represents a mythological Eagle, doing double duty as
both the traditional carrier of Zeus's thunderbolts, and a form
taken on by the King of the Gods himself. Famous for his roving
eye and relaxed approach to issues of consent, Zeus kidnapped a
beautiful Trojan boy called Ganymede, taking him off to Mount
Olympus where he was granted eternal youth and became the
official Cup Bearer to the Gods (because that was a thing,
apparently). Ganymede himself is depicted in the neighbouring
constellation of Aquarius to the east.

Aquila flies north up the Milky Way, showing its tail
feathers to the bright star clouds of Sagittarius behind it, while
apparently set for a beak-to-beak collision with the southbound
swan Cygnus further ahead. The Eagle's head is marked by
the magnitude-0.8 star Altair, and easy to spot due to both its
brightness and the distinctive fainter stars that guard its western
and eastern flanks. Altair also forms the narrow southern tip of

---

* You might recall King C as the mythical Ethiopian ruler whose awesome parenting
skills were showcased in the Perseus legend during our visit to Algol.

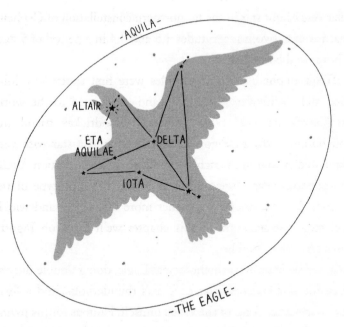

the "Summer Triangle" – a wedge of northern-hemisphere sky whose other corners are marked by Deneb in Cygnus, and Vega in Lyra, the Lyre.

The Eagle's wings spread out to either side of magnitude 3.4 Delta Aquliae, with their tips marked by the slightly brighter Theta to the east and Zeta to the northwest. Eta (sorry about all these Greek letters) lies pretty much halfway along a line between Delta and Theta, and (just like Delta Cephei) it oscillates in brightness between magnitudes 3.5 and 4.4. One easy way of tracking its relative brightness is to see whether it's better matched with nearby Delta, or with mag-4.4 Iota Aquilae, a couple of degrees to its southwest.

Pigott's initial observations included a more painstaking version of this comparison, which allowed him to show that

Eta Aquilae's oscillations follow a precise repeating period*. He estimated this to be 7 days, 4 hours and 30 minutes (which turns out to be about a quarter of an hour too much).

Throughout the nineteenth century, many more stars joined Eta and its Delta Cephei in this new class of short-period oscillating stars. Most followed the "ski-lift" pattern set by the two prototypes, with a steep ascent in brightness followed by a longer, shallower decline (and sometimes, as with Eta, a brief recovery to a second bright peak). Their periods ranged from days down to mere hours, while a distinct group towards the longer-period end of things rose and fell in a smoother, more even cycle. These became known as Geminids (named after the star Zeta Geminorum in, you guessed it, the constellation Gemini).

The breakthrough that turned Cepheids from just another astrophysical puzzle into an invaluable part of every astronomer's toolbox was made, perhaps unsurprisingly, at the hive of activity that was Harvard College Observatory. One of the many projects handled by Edward Pickering's industrious team of female computers was the analysis of photographic plates gathered from the observatory's southern-hemisphere outposts (of which more anon). In 1904, Pickering asked Henrietta Swan Leavitt, a serious-minded and assiduous veteran from the early days of the Henry Draper Catalogue, to find and analyse variable stars among the plates being sent back from Peru. Leavitt focussed, in particular, on the Small Magellanic Cloud (SMC to its friends), a misshapen tangle of stars and gas in the

---

* Interestingly, Pigott's paper refers to Eta as Eta Antinoi, placing the star in the now-discarded constellation of Antinous, named after Emperor Hadrian's best beloved, another clean-limbed Mediterranean lad.

far southern sky that appeared for all the world like a detached clump of the Milky Way.

Leavitt had already made a name for herself as a specialist in variables – despite weak eyesight and other health woes, she had a sharp mind and a talent for spotting patterns and relationships. So it was that by tracking the wax and wane of some 992 stars across many different photographic plates, she was able to pick out 16 that showed the distinctive sharp rise and slow decline over several days that astronomers recognize as the fingerprint of Cepheids. Reporting her discoveries in 1908[3], Leavitt dropped what would prove to be a scientific bombshell with her passing remark that "It is worthy of notice that... the brighter variables have longer periods."

Why exactly should that matter? Well if you recall our look at Alcyone, you may remember Ejnar Hertzsprung's neat trick of assuming that, since the stars of the Pleiades cluster are all at about the same distance, their relative brightnesses are a pretty accurate reflection of their true luminosities. Leavitt saw that the same thing applied to the SMC as a whole, and if the relative brightness of each SMC Cepheid was related to its period, then perhaps you could eventually find a way of using the period of *any* Cepheid to estimate its true luminosity.

In astronomical terms, this was a Big Deal. An independent method for measuring the luminosity of any star means you can use it as a "standard candle" – an object with a known light output, whose brightness measured from Earth reveals its true distance. At the time, our map of the Universe was still limited to a handful of nearby stars with directly measured parallax, and a bunch of statistical shorcuts that provided rough distances to objects that were further afield. A means of determining the luminosity of

Cepheids – relatively bright, easily identified stars that seemed to be scattered across space, might finally give some scale to the wider Universe.

It took four more years and precise light curves for nine more stars to establish the suspected link beyond doubt. The result was a short paper of 1912, *Harvard College Observatory Circular* 173, in which Pickering outlined Leavitt's work establishing the link between period and luminosity as no mere vague pattern, but a precise mathematical relationship[4]. The paper closed with suggestions for taking the work further – perhaps by establishing the distance to some more nearby Cepheids[*].

Hertzsprung was on the case almost immediately. Using the rule of thumb that a star's "proper" motion on the celestial sphere is likely to be larger if the star is closer to Earth, he calculated rough distances to 13 relatively nearby stars "of the Delta Cephei type". This allowed him to put some meat on the bones of Leavitt's period-luminosity relationship: he figured out that a Cepheid with a period of 6.6 days pumped out on average about 640 times more light than the Sun, and therefore appeared 7 magnitudes brighter than our own star would from a similar distance[5]. From this, he determined that the distance to the SMC stars measured by Leavitt must be about 30,000 light years[†].

The road seemed open to calculating distances for short-period variables anywhere they could be found – but of course there were a few bumps along the way. One was the recognition that the variables with really short periods (typically found in

---

[*] By convention, Pickering authored the *Circulars* as director of the observatory – but he did credit Leavitt in his opening lines.

[†] The paper (published in German) actually states *3,000 lichtjahren* – but the intention is clear so it's either a slip of the pen or a typo.

dense ball-shaped star clusters and now known as "RR Lyrae" stars) were not following quite the same relationship as the longer ones. Hertzsprung correctly suggested as early as 1913 that the period-luminosity relationship only really worked for the longer period Eta Aquilae-type variables, but it took some time to disentangle the two classes, which is one reason why Hertzsprung's initial distance estimate for the SMC was about six times too close. We'll find out more about the faster RR Lyrae stars in our next chapter.

Another problem was the question of extinction – the way that starlight is absorbed or scattered by interstellar dust over long distances, diminishing a star's apparent brightness. This matters because Eta Aquilae and its siblings are both rare *and* bright – we can see them over long distances, but even if we get better at measuring those distances, the period-luminosity relationship will be wrong if our estimates of their brightness are off.

In 1930, Swiss-born Robert J Trumpler, at the University of California's Lick Observatory, showed that dust diminishes the light of remote stars in the plane of the Milky Way by an alarming 80 per cent (1.8 magnitudes) for every 3,000-odd light years. Today's astronomers have a variety of clever tricks up their sleeves to take account of this and get a more accurate measurement of the extinction suffered by specific stars, but the quest to refine the period- luminosity relationship still continues even today.

* * *

We'll see where exactly the Cepheids led in the remaining chapters of this book, but what exactly *are* these incredibly useful stars? In terms of colour they belong to spectral classes

A and F, meaning they mostly appear white or yellow-white, with surfaces anything from a few hundred to a thousand or so degrees hotter than our Sun. Yet, as Hertzsprung found from his proper motion estimates, they are considerably brighter than the Sun, placing them above the main sequence of stellar evolution in a region inhabited by stars known as yellow supergiants.

Mass-wise, the Cepheids range between about 4 and 20 Suns, and astronomers who have untangled the story of their evolution reckon this is just the right range to send them down a unique evolutionary pathway. You may remember from our visit to Mira that many stars become red giants twice – once after they exhaust the hydrogen fuel in their cores, and once after they have burnt through all their core helium. In between, during the phase when an active helium-burning core is surrounded by a shell of hydrogen fusion, they stabilise for a short while, with their overall energy output dropping and their outer layers becoming more compact and therefore heating up. Stars

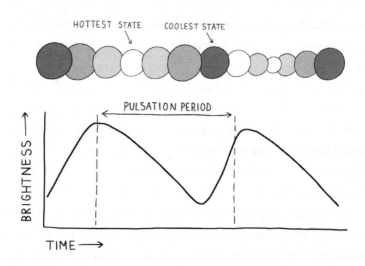

in this phase of their lives are said to be on the "horizontal branch" of the H-R diagram of stellar evolution – they drift eastward and a little south of where the red giants hang out, before retracing their steps west-northwest as their core helium is depleted (see the diagram on page 163).

If a star's mass is in just the right range, then its trek back and forth across the diagram takes it through a danger zone known as the instability strip, which might sound like something out of *Star Trek* but is notably Romulan-free. Instead, the strip marks a zone where stars can fall victim to the (equally cheesy-sounding) kappa mechanism – a kind of self-regulating stellar valve that can cause a star to expand and contract with a short but regular period.

Much of the work to put the period-luminosity relationship on a firm footing was done by a young Princeton postgrad called Harlow Shapley, who we'll be seeing a lot more of in the next three chapters. By 1914, studies of Cepheid spectra had shown that these stars were changing colour as well as brightness – shifting their light slightly towards the hot blue end of the spectrum as their luminosity increased, then growing cooler and redder as they faded. While some astronomers were determined to explain Cepheids as a type of spectroscopic binary (see Mizar), the colour change convinced Shapley that Cepheids were "pulsators" – stars whose changes in brightness and colour were linked to changes in their size[6]. As with many of our ideas about the internal structure of stars, however, working out the details of the theory was left to Arthur Eddington in the 1920s.

After dismissing the possibility that Shapley's pulsations were driven by rapid changes to the star's actual energy production, Eddington realised the effect could be produced if a specific layer was changing its opacity depending on the amount of radiation it let through (denoted by κ, the Greek letter kappa). This hypothetical layer could work a bit like a safety valve: when the star was compressed, it would turn opaque, slowing the escape of radiation from the interior, and increasing the radiation pressure from beneath until eventually it was pushed outwards. As conditions became less dense, it would then turn transparent, allowing the star to let off excess energy. As the pressure from beneath fell away, the layers would fall inwards again, turn opaque, and... well you get the idea[7].

Ingenious though Eddington's idea was, matching it up to reality proved to be a struggle, since most gases become more transparent, rather than opaque, at higher pressures. It wasn't until the late 1940s that Russia's Sergei Zhevakin pinned down a mechanism capable of doing the job. Just below the star's visible surface, he explained, are regions called partial ionization zones, where it isn't quite hot enough to completely ionize (break up) atoms of hydrogen and helium into the usual soup of electrons and atomic nuclei. As a result, there are fewer particles around to block the passage of light. However, compressing these zones encourages ionization, causing the density of particles to rise, so that light has a harder job getting through. Eddington had modeled this idea but dismissed it as unlikely based on what was then known about the composition of stars – Zhevakin showed that if there was enough helium present (about 15 per cent of the star's atoms), then the mechanism could work[8].

This explains why the instability strip sits where it does on the map of stellar evolution – a star has to reach a certain level of maturity for that much helium to be floating around inside it, and even then, the kappa mechanism can only take effect when conditions in a star's upper layers are, like Goldilocks's porridge, "just right'".

Question at the back – we still haven't explained just *why* the period and luminosity are linked so neatly? Well, this comes down to another aspect of Eddington's valve theory: as the critical layer repeatedly switches off and on and the star swells and contracts, it builds up a resonance in much the same way as a ringing bell or a vibrating violin string – a vast star-sized oscillation whose fundamental wavelength is determined by the diameter of the star (the length of the string, if you prefer).

The frequency of this vibration (and hence the period of the star's pulsations) depends on both its wavelength, and the speed at which it can propagate through the star, which is a matter of internal density. If you've been paying attention in previous chapters, you'll hopefully have a feel by now for why these factors are both ultimately dependent on the rate at which the star is pumping out energy, which will save us all a lot of painful equations[†].

Henrietta Leavitt died in 1921 at the age of 53, before the Cepheid revolution she had unleashed could gather momentum. Four years later, as we'll see when we come to Andromeda, Edwin Hubble would use her discovery to turn the Universe on its head.

---

* This doesn't mean that only Cepheids can cross the instability strip, by the way – other stars can find their own paths to a similar delicate balance of ingredients and conditions, leading to a variety of different types of pulsating star, and in fact we'll be meeting some of these in the next chapter.

† Taking the musical model further, it's possible to fit other wavelength/frequency combinations, or "overtones" into a star, just as you can produce them on a string. Interactions between a star's fundamental frequency and these overtones can explain variations on the general Cepheid theme, such as the small recoveries shown during the fading phase of longer-period Cepheids like Eta Aquilae, and the more even rise and fall of the still-slower Geminids.

20 – Imposter #1: Omega Centauri

Globular clusters: stellar cities with ageing populations

※

Scan across the skies around the richest, brightest parts of the Milky Way with a pair of binoculars, and you may spot something unusual – the occasional object that looks at first glance like a star, but appears strangely out of focus when you concentrate on it.

You might perhaps think you've found a comet – one of those small chunks of rock and ice that occasionally escapes from the deep freeze of the outer solar system and plunges towards the Sun, wrapping itself in a halo of gas that evaporates from the comet's surface as it begins to warm up from its long, cold sleep. But comets move against the background stars from night to night, while these blurry "stars" remain resolutely fixed in place. And zooming in on them with a small telescope starts to reveal the truth, as their hazy outer layers sharpen to a cloud of countless points of light: these cosmic fuzzballs are in fact vast spherical clouds of stars, far more densely packed than loose "open" clusters like the Pleiades, and home to a very different stellar population. They are called globular clusters.

The brightest globular cluster of them all was mistaken for a star for over 1,500 years. It lies in the constellation of Centaurus (which you can probably guess represents a celestial centaur), and was described by Ptolemy of Alexandria as the star on the front shoulder. When Johann Bayer put together his *Uranometria* star atlas in 1603, he gave it the designation Omega, and the name Omega Centauri has stuck.

You can't really blame Bayer for missing the fact that Omega wasn't a star – juggling his stargazing activities with a successful legal career in the Bavarian city of Augsburg, he could never have seen this wonder of the deep southern skies for himself. Instead, his atlas built on the work of others including Danish astronomer Tycho Brahe a generation before. Bayer's chief innovations were the neat (if somewhat erratic) Greek lettering system, and an expansion in the number of stars catalogued. He also popularized a dozen new far southern constellations invented by Dutch navigator Pieter Dirkszoon Keyser a few years previously.

Omega Centauri remains sadly out of view to anyone above about 40° North, but for those in other parts of the world, it is best seen in evening skies between around April and September. As with most of Ptolemy's ancient constellations, Centaurus is built to be seen "the right way up" from the Northern Hemisphere, with brilliant Alpha Centauri (*alias* Rigil Kentaurus) and Beta (aka Hadar) marking his front hooves. Ptolemy regarded the stars of the Southern Cross as part of Centaurus, though from the way they dangle beneath the figure's belly (in a manner that will be instantly familiar to anyone whose spent time around horses) we should perhaps be thankful it's now a separate constellation*. Omega, meanwhile, sits squarely in the middle of Centaurus, looking for all the world like an innocent magnitude 3.9 star.

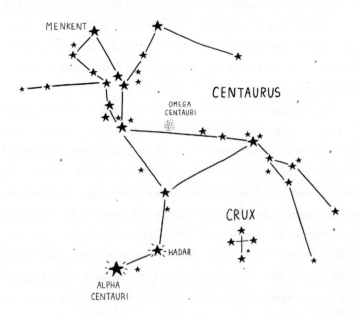

Grab a pair of binoculars, though, and that changes instantly: Omega reveals itself as a cloud of stars roughly the size of the Full Moon, loose at the edges and perhaps just slightly oval, but increasingly packed towards the centre where stars jostle together like the mosh pit at a pre-COVID rock concert.

---

* Perceptive readers may be wondering how Alexandria-based Ptolemy got a good view of the hooves and Southern Cross, as well as some other notable ancient constellations that we now consider part of the far southern sky. The explanation lies in the same precession effect that causes Earth's celestial poles to slowly wander asthe Moon's gravity tugs on the equator. As a result, previous generations of stargazers were able to view different areas of sky to those we experience today.

The first person on record as noticing that Omega wasn't a normal star was Edmond Halley, who gave it a once-over during his 1677 sojourn on the South Atlantic island of Saint Helena[1]. However, its true nature was only established in 1826 by James Dunlop, a young Scottish astronomer who had found himself working at the other end of the world after his patron, army officer and keen stargazer Sir Thomas Brisbane, was appointed governor of New South Wales. Brisbane had set up the first proper Australian observatory at Parramatta (now part of Sydney's urban sprawl), and tasked Dunlop and German astronomer Carl Rümker with cataloguing the southern stars in detail. When he got to Omega, Dunlop immediately realised that it belonged in the same family as other near-spherical star clouds known from northern skies, for which William Herschel had coined the term "globular clusters" in 1789.

Astronomers recognized at once that Omega was a cut above every other globular cluster in the sky, but the first attempts to really get the measure of it did not come until late in the nineteenth century, when Harvard College Observatory established a southern outpost in Peru under the leadership of Solon Irving Bailey. The station, funded by a bequest from Massachusetts inventor Uriah Boyden, eventually settled on a high plateau near the city of Arequipa, where Bailey – a man whose occasional social reserve was set in stark contrast to a daredevil appetite for adventure[*] – began to map the sky with a state-of-the-art 13-inch telescope, spectroscope and camera.

---

[*] When he wasn't busy at the telescope, Bailey was tasked with setting up a series of weather stations across Peru, which often involved hair-raising climbs to inaccessible mountain peaks.

Plates taken at Boyden Station were shipped back to Harvard, where they were subjected to the usual rigorous analysis by the industrious team of female "computers".

Bailey became captivated by Omega Centauri after an early viewing in 1893, and, alongside other globular clusters, it became the focus of his career. Since no individual star in the cluster was brighter than magnitude 8, he reasoned that it must be very far away and so all of its stars could be treated as lying at the same distance. This meant that differences in their apparent magnitude would reflect differences in their true luminosity, which might reveal some interesting patterns.

Bailey's photographs of Omega allowed him to pick out thousands of individual stars from the general melee, and make the first attempts at an estimate of how many stars it contained. Regular snapping also revealed that many of these stars varied in brightness with periods of around half a day. The changing stars followed a very distinctive light curve, with a steep rise in brightness followed by a slow descent and a final sharper dip before the cycle repeated. Known for a long time simply as "cluster variables", these white and yellow stars were initially assumed to simply be shorter-period versions of the Cepheids we met in the previous chapter.

By 1902, Bailey had identified an astonishing 500-odd variables in Omega Centauri alone, and published detailed measurements for the brightness range and period of some 128 of them (needless to say, the Harvard computers – in this case the indefatigable Williamina Fleming and her colleague Evelyn Leland – were doing much of the hard graft behind the scenes).[2] Later, after returning to the US, he discovered equally impressive numbers of variables in other globular clusters such as Messier

3 (in the northern constellation of Canes Venatici, the Hunting Dogs) and Messier 15 in Pegasus.

In 1914, Bailey waxed lyrical to a young Harlow Shapley on the potential secrets to be learned from the vast numbers of directly comparable stars within globular clusters. The up-and-coming Shapley – recently appointed Junior Astronomer at California's Mount Wilson Observatory – was inspired to begin his own studies, which soon yielded impressive results.

Henrietta Leavitt and Ejnar Hertzsprung's work on the Cepheids had just revealed that for this class of star, there was a clear relationship between their average brightness and the period of variability. Working on the understandable assumption that the cluster variables were also Cepheids, Shapley realised that their very similar periods must mean they had very similar intrinsic luminosities. This meant that he could guesstimate the relative distances of different globular clusters simply by measuring the apparent magnitude of their cluster variables.

The next step was clearly to find a means of turning the relative distances into actual numbers, and here Shapley thought he had struck lucky, since several of the globular clusters (Omega included) also contained variables with longer-period Cepheid-type variations. Using Hertzsprung's figures for the relationship between period and luminosity, Shapley calculated distances for the clusters that ranged between from 20,000 light years for the closest, such as Omega Centauri, to almost 200,000 years for the most distant. When he plotted the locations in space for dozens of clusters, he noticed that they seemed to congregate around a particularly bright part of the Milky Way in the direction of Sagittarius.[3]

We'll see the importance of *that* discovery when we visit the heart of our galaxy in the next chapter, but for the moment we should point out that many of Shapley's figures proved to be considerable exaggerations while the clusters do indeed form a halo around the Milky Way, Omega is actually around 15,800 light years away, while the most distant globular clusters, such as NGC 6229, are about 100,000 light years away (rather than Shapley's 150,000-plus).

Alternative: A northern glitterball

Omega Centauri may be off limits for those of us in more temperate northern latitudes, but fortunately there's a fine alternative near at hand in the form of Messier 13, the Hercules Globular Cluster. Although not quite as impressive as Omega (it contains "only" a few hundred thousand stars and lies a little further away at about 22,500 light years), at magnitude 5.8, M13 is still bright enough to find with the naked eye under dark skies. And it's an easy spot for binoculars or a small telescope, appearing as a blurry ball of light about two-thirds the diameter of the Full Moon.

To track it down, you won't be surprised to learn you need to look for the constellation of Hercules, which drifts into eastern evening skies around March and soars high overhead on northern summer nights (when many Southern-Hemisphere skywatchers can spot it looking north). The figure of the legendary hero is a bit puny to be honest, with four straggly "limbs" of stars emerging from the corners of a boxy body called the Keystone. This quadrilateral of moderately bright stars stands out under dark skies and make M13 easy to track down – you'll find it slightly more than halfway up the Keystone's western edge, between the stars Eta and Zeta Herculis.

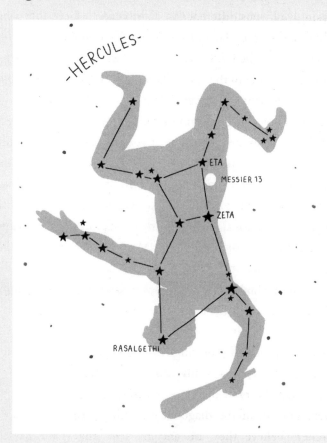

The problem for Shapley's calculations turned out to lie in the fact that cluster variables aren't really Cepheids at all – just cousins exhibiting deceptively similar behaviour. The distinction between the two groups emerged over the next decade or so: as more stars of each type were found, it became clear that while the cluster variables had periods measured in hours, there was pronounced gap between them and the shortest-period Cepheids with periods of a few days. Alongside this was an increasingly obvious difference in distribution – Cepheids were mostly found in the plane of the Milky Way, while the cluster variables (soon called RR Lyrae stars, after a relatively nearby and bright example in the constellation of the Lyre) were found *mostly* in the outlying globular clusters.

Compounding this problem was the fact that even the longer-period Cepheids in the globular clusters don't behave in quite the same way as those in the rest of the galaxy. The difference comes down to the proportions of heavy elements contained in the two types of star; even though both are driven by the same mechanism, the metal-rich "classical Cepheids" found across the galaxy shine more brightly than their metal-poor lookalikes in the globular clusters*. Assuming they must all be the same led Shapley to overestimate the luminosity of the cluster Cepheids and therefore conclude they were further away than they actually are.

The RR Lyrae stars, meanwhile, turn out to be a completely independent group of pulsating stars, with much lower masses than Cepheids (and typically only around half the mass of the Sun). On the H-R diagram of stellar evolution, they sit at the very bottom of the diagonal instability strip, and today's researchers believe they are ancient stars that have passed

---

* "Metal" in this context meaning any element heavier than helium – a loose astronomical shorthand that would give any true chemist nightmares. The presence of metals opens up the pathway to the high-speed CNO fusion cycle that allows higher-mass stars to shine more brightly.

through their red-giant phase and ended up, more or less by coincidence, on a similar see-saw between internal pressure and temperature to the Cepheids.

One thing the RR Lyraes and the cluster Cepheids share is the fact that they are very low in heavy elements. Spectroscopic studies of other individual globular cluster stars (first achieved for the bright and relatively nearby Omega) reveal a similar makeup. So why should this be?

A clue comes from the fact that globular clusters contain almost nothing but stars (with perhaps the occasional black hole thrown in) – there's none of the dust and gas required for fresh star formation. You may remember from our crash course in stellar evolution that heavier stars are like Aesop's hare, racing through the stages of their life towards an early death. This means that if you cut a star cluster off from fresh star-forming material, its higher-mass stars will gradually disappear (going out either with the bang of a supernova, or the relative whimper of a red giant), leaving only the more sedate, lightweight stellar tortoises behind to shine for billions of years.

Today's astronomers think this is just what happened early in the history of the globular clusters, perhaps 10 billion years ago or more. They started off as outsized versions of today's open clusters, created when collisions between young proto-galaxies slammed together vast clouds of star-forming gas and flung them into remote orbits. The resulting "super star clusters" contained some truly enormous stars, which the comparative lack of metals at the time* allowed to shine for longer than they would today. After a few million years, however these monster hares turned

---

* Since the gas had been less enriched by heavy elements from earlier stellar generations.

supernova, generating powerful shockwaves that blasted the surrounding gas out of the clusters altogether. Once new star formation was abruptly cut off, the remaining stars were left to slowly age, with morbid stellar obesity gradually taking its toll among the brighter and heavier ones until only those less massive than the Sun survive to the present day, plodding steadily through their fuel supply as yellow, orange and red dwarfs[*].

Omega Centauri is the poster child for this sort of system – the largest and most spectacular of the 150-odd globular clusters known to orbit the Milky Way. While most globular clusters top out at a few hundred thousand stars, Omega has an estimated 10 million, spread across a ball some 170 light years across. With a total mass of 4 million Suns, Omega is a lumbering elephant compared to its more sprightly globular neighbours – so big in fact that some astronomers think it once formed the core of an entire dwarf galaxy, whose loose outer fringes have been stripped away by the Milky Way's gravity.

At their centres, Omega and other globulars are packed so tightly that stars are separated by a matter of mere light days or even light hours. In this crowded environment, close encounters and even collisions between stars are frequent. Direct collisions, or the forging of new binary systems among the closely crammed stars, can explain one of the most enduring globular mysteries – the presence of hot, bright stars known as blue stragglers, which shine out like emergency lights among their fainter red and yellow neighbours.

The first of these stars was discovered by American astronomer

---

[*] This is the straightforward "one-and-done" version, anyway – a handful of globular clusters show evidence of two or more stellar generations, probably due encounters with independent clouds of star-forming gas after their initial formation.

Allan Sandage in 1953[4], and initially it seemed that these bright blue interlopers would spell trouble for the neat theory of globular clusters as stellar retirement communities. Burning through their fuel at an accelerated rate, they must surely have shorter lifespans and have formed fairly recently? But studies of individual stars in Omega's core since the 1990s offered a way out of this apparent paradox. They showed that not only do blue stragglers weigh up to twice as much as their neighbours, but these new kids on the block also spin significantly faster – a sure sign that our old friend the conservation of angular momentum has been at work. The stragglers are now thought to form when low-mass stars in the closely packed core regions come together in slow-motion encounters, forming unstable binary systems. Within these systems one star's gravity may drag material away from the other in an act of cosmic cannibalism, or both stars may spiral inwards until they eventually merge together. In either event, the result is a more massive star with a faster spin, increased energy output and a hotter surface.

Before we leave Omega Centauri and the realm of the globular clusters behind, there's one obvious question left to address: just why should large groups of ageing stars create such perfectly spherical clouds? Well, the answer lies in the fact that the stars' distribution at any one moment is a brief snapshot of their ever-changing motion within the cluster: in reality, each star follows an elongated elliptical orbit, moving fastest when close to the shared centre of mass*, and much more slowly at the outer limits of their orbit. Countless stars with similar orbits overlapping one another give the impression of a spherical cloud.

---

* Which may or may not be a 10,000-solar-mass black hole, depending on which group of researchers you talk to…

And how do the stars get onto these elongated orbits in the first place? Once again, we have to look towards the crowded centre, where collisions and captures may be relatively rare, but near misses happen with alarming regularity. During such close encounters, stars can swing past each other, briefly catching hands through the pull of their mutual gravity before letting go to be flung away from the core. With no gas or dust present (the sort of material that might form a flattened disc through collisions, and help to control the movement of these stellar escapees), the directions of these new deflected orbits can be entirely random. The result of this chaos, seen from thousands of light years away, is stunning.

21 – S2

A journey to the centre of the galaxy

✳

According to a 2016 study, the pollution of our night skies with light from human activity has become so bad that one-third of all people on the planet cannot see the Milky Way from where they live. In Europe this rises to 60 per cent and in North America 80 per cent, while the entire populations of small countries such as Malta, Singapore and Kuwait don't stand a chance.

Fortunately, for most of us town mice, dark (or at least dark enough) skies are still usually somewhere within reach, even if we can no longer just step out of our doors and see it overhead. Make the effort to get away from street lights, neon signs and the glare of our omnipresent screens, and the Milky Way is still there, waiting for us. Almost every ancient culture has its own folktale or myth to explain what it is and how it got there – whether it's the glowing wake of a Maori canoe, embers thrown into the sky by a Khoisan girl seeking to light up the night, or the magical pathway used by birds migrating from the uttermost north of

Finland. Classical Greek mythology, meanwhile, delivers yet another problematic tale of marital infidelity from Mount Olympus. In this one, the Milky Way is a fountain of milk from the breast of Zeus's long-suffering wife Hera, spurted into the sky as the enraged goddess awoke to find a strange infant, the baby Hercules, suckling at her breast.*

In reality, of course, the Milky Way is something far more impressive than any human mythology can come up with – a band of countless stars, stretching away to invisibility in places and marking the central plane of the vast spiral galaxy that is home to our solar system and every star we can see with the naked eye. At its northernmost extreme, it passes through Cassiopeia and Cepheus, while at its southern end it loops through Centaurus, Crux and Carina.

The best times to see the Milky Way come when it cuts roughly north-south across the sky, passing near the zenith (the point on the celestial sphere directly over our heads). For evening stargazers, that's usually around the beginning of September and the start of March, but there's a big difference between the two. While the March Milky Way puts on a rather wan show (especially for northern skywatchers) as it threads its way through the constellations near the horns of Taurus, September's apparition is brighter and more complex in general, reaching its most intense in the crowded star fields of Sagittarius, the Archer.

The reason for this difference lies in the geometry of the Milky Way and our relationship to it. Our galaxy is a wide but relatively

---

* Zeus had decided to give little Herc, his bastard lovechild with the mortal Alcmene, a crafty dose of Hera's goddessy goodness as she slept. The word *galaxy* comes from the same root as *lactose*, and in this case you can hardly blame Hera for being somewhat intolerant.

thin disc, so when we look in some directions, stars in our local neighbourhood seem well spaced out and we can easily see past them into the darkness of surrounding space. In other directions, however, stars crowd together behind each other, creating clouds of faint light as we look across the plane of the Milky Way. Just how many stars we see depends on how much of this plane we're looking through – towards Taurus, we're facing the outer edge of the galaxy and the stars are rather thin. Towards Sagittarius, we're staring straight into the heart of the matter, towards the centre of the galaxy some 26,000 light years away where a group of heavyweight stars orbit around a slumbering, unpredictable monster – the supermassive black hole that anchors the entire Milky Way.

The star known as S2 is the most intensely studied of these central stars, if no longer the star closest to the black hole itself. Unfortunately, embedded as it is in the midst of the Milky Way's central hub (a vast rugby ball of older red and yellow stars some 20,000 light years across), and with the intervening Scutum-Centaurus spiral arm (a long, diffuse collection of stars, gas and dust) also getting in the way, there's no chance for us enthusiastic amateurs to see this central region for ourselves. Instead, the best we can hope for is to know we're looking in the right direction.

So, Sagittarius: usually depicted as a centaur armed with a bow and arrow, to modern eyes this constellation looks like nothing so much as a teapot, with the brightest parts of the Milky Way forming the steam that rises from its spout*. The tip of the spout is marked by Alnasi, a magnitude-3.6 star also known as Gamma-2 Sagittarii. Follow a line from Alnasi heading a little north of west, towards magnitude-3.2 Theta Ophiuchi. About halfway along you should spot a star called 3 Sagittarii shining at magnitude

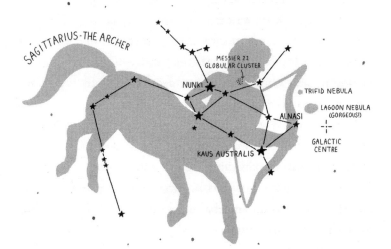

---

* With both Sagittarius and Centaurus to cope with, the night sky seems to have more than its fair share of mythical horse-men. The two constellations are generally agreed to represent the specific centaurs Chiron (tutor to a slew of Greek heroes and demigods) and Pholus, but no two authorities can quite agree on which is which.

4.6; the location of S2, and the central black hole, is about two Moon-widths to its south.

That's the centre of the Milky Way, then – but how did we get here?

★ ★ ★

The first person on record as drawing conclusions about our galaxy's physical structure from its appearance in the sky was Durham-born astronomer and instrument-maker Thomas Wright, who in 1750 suggested that the Milky Way exhibited what he termed a grindstone structure[1]. A generation later, William and Caroline Herschel, riding high from William's discovery of Uranus and appointment as King's Astronomer*, collected concrete evidence for this idea when they began an insanely ambitious survey of the heavens from their new observatory at Datchet near Windsor.

The Herschels' scheme was based on the assumptions (not unreasonable for the time) that stars were all of roughly the same luminosity and were scattered evenly across space. As a result, their apparent brightness would fade with their distance, and the number you could see in any one direction (assuming you stuck to the same telescope) would indicate how far it was to the "edge" of the Universe. By dividing the sky into 700 equal areas and counting the number of stars in each, they produced a map of the Milky Way as an amoeba-like flattened blob with the Sun somewhere near its centre[2].

The idea that the galaxy might actually have a pinwheel structure began to take hold in the later nineteenth century, after

---

* A less academic and more courtly role than the Astronomer Royal.

the discovery of spiral patterns in some distant nebulae. At the time, the nature of the nebulae was still hotly disputed (as we'll see shortly when we come to Andromeda) and the notion of a spiral Milky Way tended broadly to go hand in hand with an acceptance that the spiral nebulae were vast "island universes" on a similar scale to the Milky Way itself, rather than a completely different type of object embedded within our galaxy.

Our own place in the Milky Way, meanwhile, snapped into focus in 1921, with another Copernican downgrade to Earth's importance at the hands of Harvard's Harlow Shapley. By mapping the distribution of the globular star clusters like Omega Centauri, which were generally accepted to be loiterers in the regions above and below the plane of the Milky Way, Shapley realized they were centred around a point thousands of light years away in the direction of Sagittarius. This, it seemed reasonable to assume, was the true galactic centre around which everything else orbited.

With our solar system abruptly relegated to an uncharted backwater at the unfashionable end of the galaxy, attempts to map our surroundings could finally make some progress. In 1927, Dutch astronomer Jan Oort duly made a key discovery: different parts of the galaxy move at different speeds.

This is obvious if you think about it – the Milky Way isn't solid, but instead hangs together because every object within it is following its own orbit due to the influence of gravity. Thus it must follow the same rules of motion that affect planets in our own Solar System, or star systems such as Mizar – in other words, the further out an object orbits, the more slowly it will move.

Oort set out to map the movement of stars in different parts of the Milky Way (both their proper motion "across" the sky, and

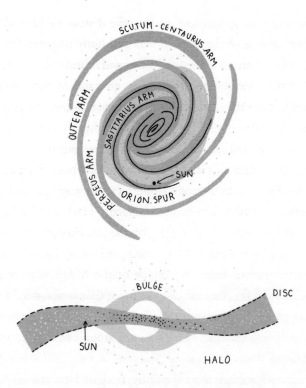

their radial motion towards or away from the solar system). Taking account of these laws of orbital motion and the Sun's own drift through space, he was able to build up the most detailed picture yet of the galaxy's structure, showing that it was a disc roughly 80,000 light years in diameter. Our Solar System lies roughly halfway from the centre to the edge of this disc (about 26,000 light years from the centre by modern measurements), and makes a full circuit around the core every 250 million years*. Thirty years later, at the dawn of the space age, Oort returned to the

---

* Modern estimates put the diameter of the Milky Way's starry disc about 20 per cent larger than this at 100,000 light years. Bigger figures are often bandied around – but it all depends on where you draw the line.

question of galactic structure, and finally established the existence of spiral arms embedded within the disc beyond doubt. He did this using the relatively new technology of radio astronomy to see through the confusion of overlapping star clouds and pin down the locations of the radio-emitting hydrogen clouds that form the Milky Way's skeleton.

We'll be returning to spiral arms and galactic rotation at the end of this chapter, but right now seems like a good place to turn our gaze back in towards the galactic centre, and what exactly is going on with the pithily monickered S2.

★ ★ ★

The radio signals from the sky that proved so useful to Oort were detected for the first time in the 1930s, when Oklahoma-born engineer Karl Jansky was tasked with investigating possible sources of interference with radio transmissions. Jansky built a steerable antenna (a Heath-Robinson device that looked more like a crop-spraying machine than a telescope) in a field near the Bell Telephone Laboratories in New Jersey, and used this to hunt down different types of radio noise. Alongside interference from thunderstorms near and far, he found himself plagued by a faint but persistent signal that came and went each day. At first, he assumed it came from the Sun, but as the months went on it drifted further and further out of step with sunrise. By 1933, he had the answer – the radio waves coincided with the visibility of the Milky Way, and specifically with the direction of Sagittarius. Something at the heart of the Milky Way, was producing powerful radio waves.

Thanks to successive generations of radio telescope dishes and the development of radio interferometry (similar to the visual methods used to resolve the detail of stars like Betelgeuse), we now have a far more detailed picture of what's going on with Milky Way FM than Jansky could have dreamed of. In the 1960s, two distinct elements were recognized: Sagittarius A East (which turns out to be an expanding supernova remnant) and Sagittarius A West, an enigmatic three-armed spiral of glowing gas. Then in 1974, Bruce Balick and Bob Brown of the US National Radio Astronomy Observatory (NRAO) found a more compact source of radio waves at the very centre of the western spiral – a mysterious object that Brown later suggested should be designated Sagittarius A* (with the suffix pronounced "A-star")[3].

★ ★ ★

Sagittarius A* marks the true heart of the Milky Way – confirmed in the early 1980s after years of careful measurement showed that, unlike everything else in the region, it doesn't shift its position[*+4]. Astronomers didn't hang around for confirmation, however, before they began to speculate about what the strange radio source might be, And from the outset, the prime candidate was a gargantuan "supermassive" black hole with a mass on the scale of a hundred thousands Suns or more.

The idea that the cores of galaxies might conceal monster black holes had developed in the 1960s from attempts to understand the strange distant galaxies called quasars (see 3C 273, coming up shortly). By 1971, British astronomers Donald Lynden-Bell and Martin Rees had suggested that

---

* * Just as stars move more slowly at greater distances from the core, the converse is also true – those in the central region tear through space at very high speeds and orbit the Sagittarius A* in a matter of decades.

what was sauce for the goose was sauce for the gander, and that perhaps all galaxies, including the Milky Way, had a central black hole acting as a gravitational anchor around which everything else orbited.

In the Milky Way's case, of course, such a black hole would have to keep as quiet as a supermassive mouse, barely letting out a peep to give away its presence and certainly not screaming out X-rays like certain black holes we could mention*. In fact with everything else in its neighborhood presumably keeping a safe distance, the monster would be more likely to reveal itself through relatively gentle radio emission, as stray particles of gas and dust drifting within its reach were dragged to their doom.

It's a nice theory, but proving it's another matter, and that (at last!) is where S2 comes in. As infrared astronomy took off in the 1980s, astronomers found they had another tool that could pierce the intervening star clouds and see what was going on around the galactic centre. Infrared images from mountaintop observatories and especially space telescopes soon revealed the unexpected presence of three brilliant star clusters in the galactic centre – the Quintuplet, the Arches, and the "S-Star cluster". The first two in particular are home to some of the most massive and luminous stars in the Milky Way, including an Eta-Carinae-like luminous blue variable called the Pistol Star that pumps out 1.6 million times more energy than the Sun†.

The S-star cluster is no match for its bigger neighbours, but its

---

* Looking at you, Cygnus X-1...

† Given the evolutionary spread of its stars, each of these clusters can only be a few million years old, so their discovery came as quite a surprise to astronomers who had assumed that the centre of the galaxy, like most of the hub region that surrounds it, would be a stellar retirement home for ageing red and yellow stars.

importance lies in the fact that it directly surrounds Sagittarius A* itself. The orbits of the stars within it can therefore reveal both the mass and diameter of whatever's producing the radio waves. Cram enough mass into a small enough space and bingo, you've got yourself a black hole.

S2 is a 15-solar-mass bluish-white member of the cluster that has been tracked continuously since 1995. (Its name, by the way, just indicates that it was the second of 11 initial cluster members discovered, counting anticlockwise around Sagittarius A*.) Regular infrared snaps of the region have captured it following an elongated orbit around the central mass with a period of just over 16 years, while measurements of the Doppler shift in its spectrum have shown that it is moving at speeds of up to 7,600 kilometres per second (38 times faster than our Solar System's own leisurely spin around the Milky Way*).

Based on early measurements of S2's movement through space, Andrea Ghez and her colleagues at UCLA were able as early as 1998 to model an orbit that came within 17 light hours of the central mass at its closest, and to show that whatever S2 is orbiting must weigh at least 2.6 million Suns[5]. While an extreme sceptic could, at a pinch, cram a monster star cluster into the available space (roughly four times the diameter of Neptune's orbit around the Sun), they'd still have to explain how such a cluster could remain stable, and be apparently invisible. This is the point at which William of Ockham's famous philosophical razor (the simplest answer is usually the correct one) comes into play and the supermassive black hole becomes the sensible

---

* In 2018, during S2's closest approach to Sagittarius A*, astronomers were able to see changes in its orbit caused by the effects of general relativity, as predicted by Karl Schwarzschild a century before.

option. The case for a black hole has only been strengthened since the millennium, with the discovery of another star, S0-102, that comes within 11 light hours of Sagittarius A* and helps push its predicted mass up to 4.3 million Suns.

The environment in which S2 moves must be a strange and violent one – for one thing it is thought to be packed with thousands of dead stellar remnants – neutron stars, white dwarfs and even smaller black holes that jostle the S-stars in their orbits. What's more, while the danger zone around the black hole has been almost entirely cleared out, it's not above the odd X-ray belch in its sleep when a small asteroid falls in. In 2019, we got some idea of its likely appearance when astronomers using the Event Horizon Telescope (a cosmic Zoom call connecting radio dishes around the world) delivered the first direct image of a black hole – the 6.5-*billion*-Sun behemoth at the centre of a distant galaxy called M87. Hopefully before too long, the EHT will be delivering something similar for Sagittarius A* itself.

* * *

With our brief stopover at the core complete, we'll shortly leave the Milky Way behind completely on an intergalactic voyage. But before we do there's one last thing to mention as we look out from our vantage point embedded within this vast cosmic pinwheel. Remember how Jan Oort figured out the geometry of the Milky Way by measuring the differential rotation of its stars? Well there's a sting in the tail.

Astronomers from Oort onwards have found that stars in the solar neighborhood and beyond rotate faster than you would expect based on a simple model. In the 1970s, Vera Rubin and Kent Ford, of Washington's Carnegie Institution (measuring the rotation of stars in remote galaxies with a sophisticated spectrograph) found that the same thing consistently applied to other spiral galaxies: stars close to the visible outer edge consistently move faster than you'd predict if the concentration of stars and other matter in galaxies reflected the overall distribution of mass within them. Instead, it seemed that large amounts of undetectable matter lie outside the visible limits of our galaxy and others − in the dark space beyond the galactic disc, and the halo region where globular clusters like Omega Centauri are found. This wasn't the first evidence of something wrong with traditional models of the Universe, but it remains one of the most convincing pointers to the existence of dark matter − the Universe's hidden, unseen and un-seeable missing mass, to which we'll return in our final chapter.

22 – Imposter #2: The Andromeda Nebula

*A neighbouring galaxy –
and the wider Universe*

So far in our travels, we've been limited to the stars of our own galaxy – even Omega Centauri, vast and distant though it is, is trapped in its own long orbit around the gravitational anchor that is the centre of the Milky Way. But our galaxy is just one of countless billions in the wider cosmos. It's time to get out our hiking boots and take our first step into a larger Universe.

While every individual star and star cluster that you can spot with your naked eye is a member of the Milky Way, the night sky holds three other exceptional objects. Two of these are the Large and Small Magellanic Clouds – shapeless masses of gas, dust and stars hidden amid the far southern constellations and first brought to widespread European attention after 1522 by survivors from Portuguese explorer Ferdinand Magellan's ill-fated circumnavigation of the globe. Though technically galaxies in their own right, the Magellanic Clouds are small ones, and probably held just as much in thrall to the Milky Way as globular clusters like Omega Centauri.

The other exception, however, is a completely different matter. The Great Andromeda Nebula, aka the Andromeda Galaxy (or more prosaically Messier 31), is easy to spot once you've got your bearings, but it's so far away that its light has travelled for 2.5 million years to be with us here tonight.

Andromeda's namesake constellation represents the ill-used princess from the legend of Perseus and Medusa (for a potted version, see Algol). Along with her retinue of related star patterns, she glides into evening skies from the east around August, and doesn't disappear into the sunset until around February (depending a little, of course, on where you live). Although not particularly bright and rather shapeless in itself, her constellation has the benefit of being joined at the hip (quite literally) to the much larger and more obvious pattern of Pegasus, the Flying Horse.

You might remember from our previous trip to Helvetios that Pegasus appears upside down for Northern-Hemisphere stargazers as it backflips above the southern horizon, while southern observers, looking northwards, get to see it the right way up. A large, mostly empty square makes up the horse's body, with a chain of stars emerging from blue-white Markab (Alpha Pegasi) in the southwestern corner to form a head and neck, and two forelimbs stretching out at a canter from the red giant Scheat in the northwest (see page 129 for a map).

Tonight, however, we're interested in the northeastern corner. Its white star, Alpheratz, is something of a curiosity – despite notionally marking the root of Pegasus's hindquarters, it's officially Andromeda's brightest star, shining at magnitude

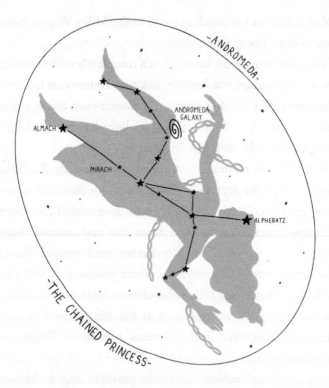

2.0 and marking the princess's head*. Provided your sky isn't entirely bleached by light pollution, you should be able to spot Andromeda's body as two branches of stars rooted at Alpheratz and curving below the obvious bright W-shape of her mother Cassiopeia. The Andromeda Nebula lies a little north of the northern chain, roughly level with the constellation's second brightest star, Mirach, in the southern one. Under dark skies it can be pretty obvious as a fuzzy cloud of light that your eye refuses to focus into a star, but there's no shame in using binoculars.

---

* Alpheratz used to be the subject of an awkward joint custody agreement between the constellations, and you can still find it sometimes referred to as Delta Pegasi.

Officially, the nebula shines at magnitude 3.4, but that's a little misleading since it's a measure of the total light spread out over an area roughly six times the size of a Full Moon. It's definitely a naked-eye object, though, which makes the fact that it was overlooked by the astronomers of the classical Mediterranean something of a puzzle. Instead, the first definitive report comes from Abd al-Rahman al-Sufi, a court astronomer to the Persian emir at the city of Isfahan in the mid-tenth century. Al-Sufi's *Book of Fixed Stars* was a translation and major upgrade of the Ptolemy's *Almagest* (which by now was 800 years old and starting to show its age) – and among the many bells and whistles he added were mentions of both the Large Magellanic Cloud (known to Arab astronomers in Yemen) and the curious "nebulous smear" in Andromeda[1].

Over the next couple of centuries, Al-Sufi's work percolated into early medieval Europe through a mix of translation, transmission and plagiarism, so that by the early 1600s, the curious cloud in Andromeda was an obvious target for the newly invented telescope*. The first person we know to have taken a look was Germany's Simon Marius (today mostly remembered for a bitter spat with Galileo about who first discovered the moons of Jupiter). In 1612, Marius noted that the nebula's light was spread across a large area of the sky but increased to a concentrated spot at the centre, like "a candle shining through horn."

While a parade of astronomy's great and good went on to catalogue Andromeda's Great Nebula, they struggled to find

---

* Top tip: Andromeda's light is so spread out that unless you have a fairly big telescope, you're likely to be far better off looking at it through binoculars with a nice wide field of view.

any detail to its structure even as telescopes improved*. This did little to discourage them from speculating about the nature of Andromeda and the other nebulae were, however, and from the mid-Eighteenth century onwards, opinion and evidence swung back and forth between two basic interpretations.

First out of the blocks was the "island universe" theory, which argued that the nebulae were incredibly distant clouds of stars – galaxies in their own right, similar to and separate from the Milky Way. Early evidence to support this came from the fact that many apparently blurry objects *did* resolve themselves into clusters of stars if you had a good enough telescope.

This early lucky guess, however, was soon eclipsed by the latest flavour of the month – a spinoff from the Kant/Laplace nebular hypothesis of solar system formation (touched on during our visit to the young stars of the Trapezium and T Tauri). Laplace's idea convinced many that the nebulae were actually sites where stars and planets were being born.

The luxury of hindsight, of course, helps us to pinpoint the problem: while astronomers of the time assumed that all nebulae must be one thing or the other, they actually include both nearby star factories *and* distant galaxies. In the Victorian era, however, support for the nebular version became so ingrained that even spectroscope wiz William Huggins, upon finding in 1864 that Andromeda's spectrum showed a star-like continuum of light rather than gas-like emission lines, resorted to excuses. By the turn of the twentieth century, the idea of external galaxies had become little more than a scientific joke.

---

* The most significant discovery came in 1749, a young French stargazer called Guillaume LeGentil spotted a blob of light a little to the south of Andromeda's nucleus. M32, as it's now known, is one of two small but bright companion galaxies in orbit around the Great Nebula, and an easy binocular spot.

So what happened to turn things around? Well, remember Andrew Ainslie Common, the Geordie plumbing engineer and telescope addict who took that stunning picture of the Orion Nebula in 1883? A couple of years later he sold his enormous 36-inch reflector to make way for a new and even bigger model. Politician and carpet magnate Edward Crossley then spent ten happy years cataloguing double stars with Common's cast-off, before finally conceding defeat to the unpredictable Halifax weather. In an act of archetypal Victorian philanthropy, he then shipped Common's redundant telescope across the Atlantic, where it became the centrepiece of the University of California's Lick Observatory.

The Lick was the world's first purpose-built mountain-top observatory, an astronomical outpost in the mountains above San Jose, where a combination of crystalline skies and a thorough refurbishment turned the Crossley Reflector (as it became known) into the Hubble Space Telescope of its day. Long-exposure photos taken by Lick director James Keeler and his successor Edward Fath were soon turning up countless new nebulae looming from the darkness of mostly empty space away from the Milky Way.

The new discoveries became known by the catch-all term "spiral nebulae" – even if quite a lot of them weren't very spiral-y at all*. With a range of shapes that varied from perfect pinwheels and ragged clumps, to cigars and ball shapes to match any sport of your choice, they displayed absorption spectra without exception, suggesting that however blurry they might look, they were in fact made up of countless stars.

---

* The discovery that at least some nebulae had spiral structures had been made in the mid-nineteenth century using the Earl of Rosse's enormous Leviathan telescope in Ireland, but this had only reinforced the infant-solar-system theory.

Photographs also revealed new structure in nebulae that were already known – most notably (as shown above) the swirls of light-absorbing dust that helped refine Andromeda's shape from an oval blur into a flattened spiral viewed from a little above edge-on. Andromeda and a few other larger nebulae even began to reveal individual stars, though supporters of the nebular hypothesis explained these away as misleading clumps of matter in a gallant rearguard action.

Two ingenious lines of research, however, soon reinvigorated the island universe theory. In 1910, affable classics-scholar-turned-astronomer Heber D. Curtis returned to the Lick from a posting in Chile and was tasked with continuing the survey of spiral nebulae with the Crossley Reflector. Sifting through his steadily accumulating collection of photographic plates, he began to notice occasional star-like points flaring to life in several of these nebulae before slowly fading away.

Curtis realised that the rise and fade of these outbursts matched those of novae (such as RS Ophiuchi and its brighter, less predictable cousins). Even with the mechanism that powered novae in the Milky Way still unknown, it seemed clear that they often peaked at around the same luminosity (a few tens of thousands of times that of the Sun). Curtis put forward a convincing case that the eruptions in the spiral nebulae were simply novae in remote and independent galaxies, seen over a great distance.[2]

Meanwhile, from around 1906, Vesto Slipher of the Lowell Observatory in Flagstaff, Arizona*, began a project to capture detailed spectra of the spiral nebulae. Although Slipher started out looking for evidence that the nebulae were rotating incipient solar systems, he soon stumbled upon something unexpected – every nebula he studied displayed spectral lines that were shifted significantly towards the blue or red compared to their expected position. After considering other possible explanations, Slipher concluded that these shifts were caused by the Doppler effect – in other words the nebulae were all in rapid motion either towards or away from Earth. Andromeda, the first to be

---

* A private observatory endowed by Bostonian businessman Percival Lowell. Despite its founder's unfortunate predilection for wild goose chases (including the cataloguing of canals on Mars and the hunt for Planet X), the Lowell became a serious research centre – perhaps best- known today for the 1930 discovery of Pluto.

successfully measured, was heading towards us at around 300 kilometres per second*, but most of the other nebulae seemed to be moving away [3]. The velocities of the nebulae were uniformly far greater than those of all but a handful of "runaway" stars (similar to those kicked out of the Orion Nebula). By 1917, Slipher had decided the most plausible interpretation was that the nebulae were island universes similar to the Milky Way, and that our galaxy was on the move in relation to them.

Slipher's discovery formed one line of argument a few years later, when Washington D.C.'s Smithsonian Museum of Natural History staged a showpiece discussion on the scale of the Universe. Subsequently known as the Great Debate, it set Heber Curtis, now Director of the University of Pittsburgh's Allegheny Observatory, up against Harlow Shapley of Mount Wilson Observatory in Pasadena [4].

Shapley's measurements of globular clusters had convinced him that the scale of the Milky Way was vast (perhaps 300,000 light years across), and that it must therefore be an all-encompassing system containing everything in the Universe. His killer line of attack, however, was furnished by Adrian van Maanen, a Dutch colleague at Mount Wilson. Van Maanen claimed to have detected visible rotation in some spiral nebulae by comparing snaps taken a couple of decades apart.

Shapley's argument was simple and inarguable: if van Maanen's estimate that the outer parts of the nebulae made a complete rotation every 10,000 years or so was correct, then they could not possibly be the Milky Way-sized objects Curtis proposed – since if they were, then stars in their outer reaches would have to

---

* Downgraded to 110 kilometres per second in recent measurements, but still pretty impressive.

be moving faster than the speed of light. Although the "debate" produced no conclusive victor, Curtis had no choice but to concede that *if* van Maanen was right, then the island universe hypothesis fell.

But van Maanen was wrong.

Four years later, the question was settled beyond doubt, thanks to the work of Missouri-born Edwin Hubble, a man who mixed undoubted scientific genius with an ability to hog the historical limelight, whether intentionally or not. Hubble joined the staff at Mount Wilson in 1919, just in time to benefit from the newly completed Hooker Telescope, a 100-inch reflector (2.5-metre) that was the largest in the world. Intrigued by Slipher's Doppler shift discoveries (and with Shapley relocated to the other side of the country to take up the directorship of the Harvard College Observatory in 1921), Hubble began a project that he hoped would settle the debate and prove that the Universe extended beyond the Milky Way.

As instruments had improved over the preceding years, several astronomers reported seeing novae and other transient stars in Andromeda and another large spiral nebula designated Messier 33 (in the constellation of Triangulum, the, erm, Triangle). Using the Hooker, Hubble identified large numbers of stars along the outer fringes of both objects, and saw that some of them changed in brightness from night to night.

The hard graft of capturing light from the nebulae onto a series of photographic plates (65 for Messier 33 and 130 for Andromeda) was largely left to Milton Humason, an unassuming high-school dropout who had started out as a janitor and quickly become a vital member of the observatory staff. Scanning through dozens of suspected variable stars,

Hubble soon spotted the unmistakable rise and fall in the brightness of Cepheid variable stars (flip back to Eta Aquilae if you need a refresher).

Hubble eventually identified a dozen Cepheids in Andromeda and 22 in Messier 33 – the game now was to nail down their cycles of brightness with sufficient accuracy to use the period–luminosity relationship discovered by Henrietta Leavitt and calibrated by Ejnar Hertzsprung. By late 1924, he had found what he was looking for – a distance for both nebulae of around 930,000 light years, putting them far beyond the limits of the Milky Way[5]. Later refinements to our definition of Cepheids (see Omega Centauri) have since boosted Andromeda's distance to 2.5 million light years, and Messier 33's to 2.7 million.

Hubble's discovery (leaked to the newspapers before it was formally announced on 1 January, 1925) catapulted him to overnight fame and cemented his place in the pantheon of *astronomers you might actually have heard of.* By settling the Great Debate once and for all, he ensured another diminution in the status of planet Earth – and we'll take up the story of what happened next shortly, when we come to our penultimate "star."

★ ★ ★

For now, though, we can't leave Andromeda behind without basking briefly in its magnificence. As galaxies go, it's a biggie, with a visible diameter of 220,000 light years, which is substantially wider than the Milky Way, and a mass that's probably about the same (though ask two extragalactic astronomers for the weight of either galaxy, and you'll end up with three different answers).

Andromeda seems to have the Milky Way soundly beaten in

terms of sheer number of stars, too – rough estimates reckon there are about a trillion, which is two or three times more than our own galaxy. In the 1940s, Walter Baade used the 100-inch Hooker Telescope to study Andromeda in detail and noticed a pronounced difference between the stars in its central bulging hub region, which were mostly red and yellow, and those in the disc, which were predominantly white and blue[6]. Drawing a comparison between the bulge stars and those found in globular clusters like Omega Centauri, he came up with the concept of two different stellar populations, defining Population I as the brighter, hotter stars found in galaxy discs, and Population II as the red and yellow stars found in globular clusters, the cores of spiral galaxies, and sometimes scattered through the halo region above and below the disc*.

Since we tend to treat Andromeda as a handy model for understanding our own galaxy and others, it's worth reiterating what's really going on from the lessons we learned in the Milky Way. The regions dominated by Population II have only ended up that way because they've been starved of new star-forming materials and their more massive and brighter stars have evolved their way out of existence. Population I stars, meanwhile, continue to thrive in the disc area, where gas and dust for new star formation are still available.

Through their gravitational influence, Andromeda and the Milky Way together control a 10-million-light-year region of space, surrounded by several dozen other galaxies in a small

---

* Baade's terminology looks a bit anthropocentric in hindsight, since we now know that Population II stars formed earlier than Population I. However, this confusion doesn't seem to have prevented some astronomers calling a hypothetical even earlier generation of stars "Population II".

cluster called the Local Group. Other members of this cosmic protection racket range from independent systems (the smaller spiral Messier 33) and substantial satellites like Messier 32 and the Magellanic Clouds, to tiny hangers-on that are little more than sparse wandering star clouds. Both of our galaxies are predators, prone to throwing their weight around, bullying the little guys and occasionally gobbling them up completely.

In around 4.5 billion years, however, the joke will be on us, as Vesto Slipher's discovery of Andromeda's motion through space comes home to roost. The Milky Way and Andromeda are heading towards each other on an inevitable collision course, and the resulting merger will tear them both apart before reassembling them into something else entirely. During the billion years of turbulence that will follow, the slumbering black holes at the heart of the coalescing galaxies may even spark to life as a source of violent activity – but if they do, it's unlikely to be with the ferocity of the next "star" on our list.

23 – IMPOSTER #3:
3C 273

Quasars – beacons in the
distant cosmos

$\star$

*E*dwin Hubble's 1925 confirmation that the Andromeda Nebula was, in fact, a galaxy beyond our own opened up an entire Universe for exploration, and our penultimate destination is among the most spectacular objects found within it. Despite a truly mindboggling distance of 2.44 billion light-years, 3C 273 sneaks onto our short list of imposters because, on first discovery, it was mistaken for a very peculiar star.

First, that rather unprepossessing name: 3C 273 simply indicates that this is object #273 in the Third Cambridge Catalogue of Radio Sources – a survey conducted using a fairly primitive radio interferometer at Cambridge in the late 1950s. For radio astronomers, interferometry (the clever-clever combination of signals that have travelled along slightly different paths to reveal hidden details, discussed on our visit to Betelgeuse) is not a luxury but a necessity. Although we often overlook it, the level of detail that a telescope can see depends not only on its size, but also on the wavelength of the waves it is trying to focus. For visible light, the difference is often negligible, but radio wavelengths are so long that even those vast dish telescopes at Jodrell Bank see the radio sky as a blur. Fortunately, in the late 1940s, Martin Ryle – a former wartime radar boffin who had gone on to work at Cambridge University's Cavendish Laboratory – invented a way of hitching together the electronic signals from separate radio telescopes to see how they compared when pointing simultaneously at the same area of the sky*.

3C 273 popped up on the survey as a "Class II" radio source – one that lay far away from the plane of the Milky Way and might therefore be a distant object in another galaxy. However, even with the use of interferometry, its location proved hard to pin down until Cyril Hazard, working at the Parkes Radio Observatory in New South Wales, took advantage of a chance 1962 alignment called a lunar occultation. This involved the Moon passing directly in front of the area of radio emission – by precisely timing the moment the signal cut out, Hazard and his colleagues were able to pinpoint its direction in space, as well as showing that it was a compact object that switched off and returned pretty much instantaneously. On account of this star-like nature, 3C 273 clearly fell into a class of objects imaginatively called "radio stars".

Halfway around the world, Maarten Schmidt, a graduate of Leiden Observatory in the Netherlands now working at the California Institute of Technology, was waiting for just this kind of information. As one of a handful of astronomers paying attention to these strange radio sources, he was keen to see if its position matched up with a visible counterpart of any sort. Using the

---

* Ryle and his colleague Anthony Hewish won the Nobel Prize for this and the subsequent discovery of pulsars, in the same year that Jocelyn Bell conspicuously didn't.

enormous 200-inch Hale telescope at CalTech's Mount Palomar Observatory (the largest functional telescope in the world for more than 40 years after its inauguration in 1948), he scanned the area, and found what appeared, at first glance, to be a 13th magnitude star. He even managed to split its faint light into a spectrum in the hope of capturing information about its chemical make-up.

Before we carry on with the story, this is probably as good a place as any to explain how and where to see 3C 273. With an average magnitude of 12.9, it's the sort of object that really needs a mid-sized telescope to see. Nevertheless, we've pointed the way to every other headliner in this book, so we're not going to let this one beat us...

Our target lies in the zodiac constellation of Virgo, the maiden. As the second largest star pattern in the entire sky, you'd think Virgo would be fairly distinctive, but to be brutally honest she's not – aside from its brightest star, Spica*, the constellation sprawls untidily across the sky in a jumble of third and fourth magnitude stars. It's a feature of evening skies around the world between February and July, and visible from November onwards for early risers.

Despite its shapelessness, Virgo is easy to spot because of Spica – a slightly variable star that averages magnitude 1.0 and is frequently joined in the sky by the nearby Moon or planets. Northern-Hemisphere skywatchers can get an additional pointer by extending the arc of the Plough's handle in Ursa Major, and imagining the curve passing through bright orange Arcturus (in Böotes, the kite-shaped Herdsman constellation) and on down

---

* Spica is usually depicted as an ear of wheat, betraying Virgo's secret past as a fertility and harvest goddess (the Greek Demeter or Roman Ceres) with a heritage stretching back to ancient Babylon at least 3,000 years ago.

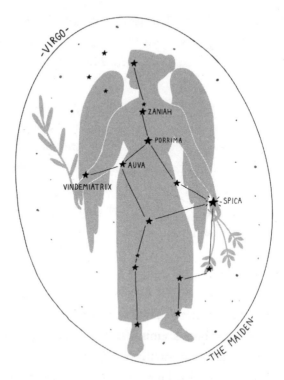

past the Celestial Equator. Southern-Hemiphere observers, meanwhile, can simply follow the long axis of Crux, the Southern Cross, in a northerly direction to reach the general area.

To find the rough location of 3C 273, you'll need to go to the other end of Virgo's body from Spica, and find Eta Virginis or Zaniah, usually imagined as somewhere along the goddess's left arm or shoulder. Our target is about 7 Moon-widths northeast of here, forming the right-angled corner of a triangle between Eta and the brighter star Porrima to the east*.

---

* To be brutally honest, even if you can spot 3C 273 you probably won't get much out of it – if you've got a telescope that's up to the task, you're better off spending time looking at the beautiful but tight double star Porrima instead.

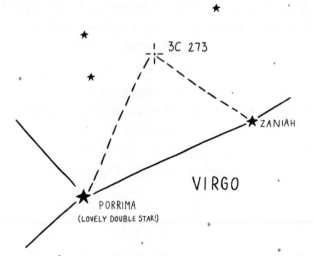

Anyway, back to Schmidt: When he looked at the spectrum of his mystery object, he found that it was unrecognizable. There were four broad emission lines, but none of their positions matched with anything known from other stars or from elements on Earth. This wasn't too much of a surprise, as Schmidt's CalTech colleague Allan Sandage had gone through this process a couple of years previously, hunting down another radio star called 3C 48. However, at this point, inspiration struck – Schmidt noticed a resemblance in the strength and spacing of the lines to a very well-known series of emissions linked to hydrogen on Earth. Calculations and cross-checks with another CalTech colleague, Bev Oke, soon confirmed the truth: these were the same elemental fingerprints, but they had somehow been shifted substantially towards the red end of the spectrum. The most likely explanation for the out-of-place lines, of course, was the Doppler effect that shifts light to the blue or red depending on whether its source is approaching or receding. But in this case, the enormous red shift implied that 3C 273 was moving away at a staggering 47,000 kilometres per second, or just over 15% of the speed of light[1].

Schmidt had discovered the first in an entirely new class of object – a distant and immensely powerful beacon of intergalactic radiation whose tremendous speed indicated a distance of billions of light years, yet which was capable of changing its brightness substantially in a matter of days. Describing the discovery in 1963, Schmidt coined the term "quasi-stellar object". Within a year, however, everyone was calling them quasars.

* * *

To fully appreciate what Schmidt had found – and why 3C 273's mindboggling speed implies an equally mindboggling distance – we need to fill in a couple of big developments that followed on from Hubble's 1925 breakthrough.

Most importantly, Hubble didn't rest on his laurels. Intrigued by Vesto Slipher's discovery of large Doppler shifts in the light of the "spiral nebulae" (indicating that they were moving towards or away from us faster than any star), he set out to discover whether there was a pattern to it. Thus far he only had two data points – the distances for Andromeda and Messier 33 – so the next few years were largely spent photographing new galaxies, tracking down their Cepheid variable stars and working out their distances.

By 1929, Hubble had used his Cepheid method to establish the distance to two dozen galaxies – confirming that not only the spiral nebulae, but also a wide range of ball-shaped "elliptical" systems, flattened "lenticular" discs and shapeless "irregular" clouds lay far beyond the Milky Way. When he plotted this new extragalactic menagerie on a graph of distance against radial velocity, he found the pattern he had hoped for. Broadly speaking (if you ignore the very nearby galaxies that display motion *towards* the Milky Way), the further away a galaxy is, the faster it is fleeing away from us.[2] And no, it isn't because we smell.

In fact, this sort of pattern had been predicted a couple of years before by a Belgian priest and astronomer called Georges Lemaître.[3] Despite (or perhaps because of) his clerical calling, Lemaître was a keen cosmologist – one of the first generation of scientists to investigate the shape and structure of the Universe itself. Their newly forged tools came in the form of Einstein's field equations – the maths that described how general relativity, the relationship between space, time, motion and mass, manifested in the Universe.

When Einstein published his equations in 1915, he had included what you might call a fudge factor. Realising that under his new rules spacetime would rapidly collapse inwards due to the mass contained within the Universe, he introduced a "cosmological constant" – a force or expansion that would prevent this from happening and allow the cosmos to survive forever (an eternal Universe being the general scientific assumption at the time).

In the 1920s, however, two scientists independently pointed out that there was another option – the Universe could avoid collapse if it was, in fact, expanding. Russian physicist Alexander Friedmann, who first floated the idea in 1922, provided equations to describe the expanding Universe scenario, but no evidence to back it up. Five years later, Lemaître reintroduced the idea, and suggested the expansion would make itself known through an increasing speed of recession for more distant galaxies – exactly what Hubble had found.

So, Hubble and others rightly interpreted the expansion not as a sign of Earth's unique unpopularity in the cosmic playground, but as evidence of a wider expanding Universe in which everything is moving away from everything else. Galaxies are dragged apart like raisins in a rising cake batter; the rate of expansion is the same for each light year of space, but objects that are separated by greater distances have more space between them to expand and therefore drift apart at higher speed*.

For an encore, Hubble turned the principle he had discovered on its head. Calculating the average rate of expansion per million light years (a figure now known as the Hubble Constant), he showed that you could guesstimate a galaxy's distance simply from the rate at which it was retreating. Later astronomers would ignore the intervening calculations and use the red shift of a galaxy's spectrum as a direct indication of its relative distance, which is why the discovery of 3C 273's huge redshift caused a bit of a hoo-hah.

We now return to your scheduled programming…

★ ★ ★

3C 273's Doppler shift implied that this strange shifting star was about 2.4 billion light years away, and when Allan Sandage reinterpreted his own pet quasar in the light of Schmidt's discovery, the suggested distance of 3C 48 came out at an even more astonishing 3.9 billion light years.

Naturally, some other astronomers had trouble believing these

---

* You may have noticed that this "cosmological" red shift is therefore not quite the same as the traditional Doppler effect, even if its effects are the same.

figures*. For a start there was the simple question of the quasar's apparent brightness – to shine at magnitude 12.9 in Earth's skies from such a great distance, 3C 273 would have to be 4 *trillion* times more luminous than the Sun. Worse still, while that sort of figure might be just about plausible for some monstrous distant galaxy, all the evidence pointed to 3C 273 being *tiny*: not only did it appear as a single bright point in the sky, but its output in a broad range of radiations varied unpredictably from hour to hour and day to day, suggesting that it couldn't be much larger than Neptune's orbit around the Sun.

What the heck could generate that sort of power in such a tiny space? As it happened, astronomers had been worrying about a similar problem (though on a less dramatic scale) for some time. Since the 1940s a number of galaxies with abnormally bright but variable, star-like points of light at their cores had been discovered. These so-called Seyfert galaxies† also showed unusual multiple emission lines, apparently emitted by glowing gas escaping from their centres at a range of speeds, with its light Doppler shifted by different amounts. Then in the early 1950s, the first radio surveys had revealed several distant galaxies surrounded by huge clouds of gently glowing gas that were apparently being released from the centre. Both Seyferts and these new "radio galaxies" seemed to have unexpected and violent activity in their cores, and when signs of faint "host galaxies" around the quasars were found, the theory

---

* One early attempt at denial – the suggestion that the quasars were actually some sort of extreme, but relatively nearby, runaway star – was neatly punctured when Dennis Sciama and Martin Rees asked, if that was the case, why there weren't similar radio stars with extreme blue shifts hurtling towards Earth.

† Named after their discoverer, Ohio-born Carl Seyfert.

that these objects were all displaying variations on the same basic activity gained ground.

Throughout the following decades, advances in radio astronomy and new satellite telescopes showed that Seyferts, radio galaxies and quasars formed a continuum – rather than being individual and sharply defined objects, one sort of galactic activity frequently went hand in hand with another. The driving force behind this activity was proposed as early as 1969 by astrophysicist Donald Lynden Bell[4], but it wasn't really taken seriously until the 1980s (and it took the Hubble Space Telescope to confirm it through observations around the turn of the millennium).

Quasars, it's now clear, are turbulent galaxies whose cumulative starlight is overwhelmed by brilliant emissions from a tiny region at their centre called an active galactic nucleus (AGN). Here, a voracious supermassive black hole, with the mass of millions or even billions of Suns, is feeding on anything that comes too close.

As large amounts of material fall towards the black hole, they are pulled apart by tidal forces and dragged into a solar-system-sized "accretion disc". Temperatures in this disc can reach millions of degrees, causing it to radiate across the electric spectrum from radio waves through visible light to X-rays. Many particles spiraling inwards across the disc find themselves caught up in the black hole's intense magnetic field. Rather than disappearing into the event horizon, they are instead accelerated to almost the speed of light before being spat out in radio-emitting jets from each magnetic pole. The disc is surrounded on its outer edge by vast clouds of opaque dust and gas, and variations in our viewing angle help determine whether a quasar is dominated by its surrounding radio emission or the visible light from its centre.

The AGN mechanism is thought to power Seyferts and radio galaxies as well as quasars like 3C 273, but a big question is just why quasars generate *so much* more energy than other types of "active galaxy". To answer that, we must tiptoe briefly into the rarefied word of cosmology.

Remember for a start that a light year isn't *just* a distance measurement. That 9.5 million *million* kilometres is the limit of how far light, the fastest thing in the Universe, can travel in a single year. Thus, light from Alpha Centauri has taken about 4.4 years to reach Earth, and we are seeing the Andromeda Galaxy as it was when its light set out 2.5 million years ago. This quirk of astronomy, called "lookback time", has little effect on the way we see the nearby Universe – what's a few centuries in the vast lifespan of a star, or even tens of millions of years in the history of a galaxy?

But things do change over time. The passing of each stellar generation enriches space with new heavy elements that become the raw materials of new stars and affect the way they in turn live and die. And as Hubble and Georges Lemaître found, on a very basic level the Universe is getting bigger and pulling galaxies apart on the largest scales.

In 1931, Lemaître asked just what all this said about the Universe of the distant past. It wasn't too difficult to figure out that the ancient Universe would have been smaller and hotter, with matter more densely packed inside it and collisions more frequent. Taking things to their logical conclusion with the help of quantum physics, the Belgian priest concluded that the entire cosmos – the fabric of space and time and everything contained within it – emerged from the explosion of a "primeval atom" with unimaginable density and temperature[5]. The idea began to find support both in theoretical physics and astronomical observations, although the opinionated Yorkshire-born astronomer and broadcaster Fred Hoyle was having none of it – in 1949, he described it somewhat dismissively as nothing but a "Big Bang". The rest, as they say, is history.

Getting back to quasars, however, one of the most conspicuous things about them is the fact that they're nowhere to be found in our nearby Universe. We only ever see them over vast distances, and this means that they're also a long way back in time. We're not quite looking back to Lemaître's primeval atom*, but we are dealing with objects whose light has travelled for many billions of years to reach us, and that's long enough for the lookback time to open a window onto a different and more violent cosmic era. Thanks to images from the Hubble Space Telescope and the enormous new ground-based telescopes built since the 1990s, we've now got a better idea of just what is going on in the space around quasars, and to be frank, it's a bit of a hot mess.

Quasars formed in an era when larger galaxies were gradually coming together from the violent collisions of smaller ones, shoveling vast amounts of material directly into their central black hole engines and fuelling the creation of solar-system-sized, blazing-hot accretion discs. Alongside them, we see intense waves of star formation, with huge monster stars living and dying in a few million years before flinging their contents out across the ancient Universe.

Since we don't find any quasars in the nearby Universe where lookback times are relatively short, we can safely assume that quasars are no longer around today. So where did they go? Why

---

* The Big Bang is now fairly well pinned down to about 13.8 billion years ago, and it took a few hundred million years after that for galaxies to begin to form.

did these cosmic dinosaurs all apparently go extinct a few billion years ago?

The truth is that, just like the dinosaurs, they're still here. But while the dinos at least had the decency to squeeze through an extinction bottleneck (with only birds emerging on the other side), quasars simply changed their name, covered up the tattoos and settled down to a quieter life. We're sitting in one at this very moment.

The absence of quasar activity in the modern Universe isn't because they're a different, alien class of object. They're a phase that every galaxy probably goes through, as confirmed by the presence of supermassive black holes at the centre of the Milky Way and many other galaxies. 3C 273 itself was a late hanger-on in the quasar game, refusing to move with the times until just a few billion years ago. Relatively nearby, present-day Seyfert and radio galaxies, meanwhile, might gamely attempt to recapture the spirit of their anarchic youth, but they're a pale shadow of their former selves.

If quasars teach us one thing about the cosmos, it's that everything changes over time. So what will the Universe look like in the future?

24 – SUPERNOVA 1994D

*Dark matter, dark energy and
the end of the everything*

---

*I*t seems fitting to bring our history of the Universe to an
end with the story of astronomy's most recent (and perhaps
still unfinished) revolution – especially as this particular
breakthrough concerns nothing less than the fate of the Universe
itself. And after our excursion to the distant realm of quasars, we
can at least begin this story a little closer to home.

In 1994, astronomers from UC Berkeley and Princeton,
using a relatively small automated survey telescope, discovered
a supernova explosion in the relatively nearby galaxy NGC
4526[1]. So far, so unremarkable, you might think – astronomers
have been tracking supernovae in distant galaxies since Walter
Baade and Fritz Zwicky in the 1930s, and the introduction of
computer-controlled automated searches has increased the
number of discoveries exponentially. But Supernova 1994D was
something out of the ordinary – not an exploding star as we know
it, but something even stranger and, for those wishing to probe
the Universe' darkest secrets, more useful.

SN 1994D, as it's known for short, faded back to obscurity long ago, so the best view we can offer today is of its host galaxy. Shining at magnitude 10.7, NGC 45526 is a fairly easy target for small telescopes, and a southern outlier of the famous Virgo galaxy cluster.

Virgo, as we may have mentioned in the previous chapter, is a rather shapeless star pattern that draws attention mostly through its brightest star, magnitude 1.0 Spica. The constellation slides into eastern skies after midnight from about February, and is at its highest in mid-evenings around May (hanging over the southern horizon for Northern-Hemisphere skywatchers, and seen in the north for those south of the equator).

Spica marks the ear of wheat the celestial maiden and harvest

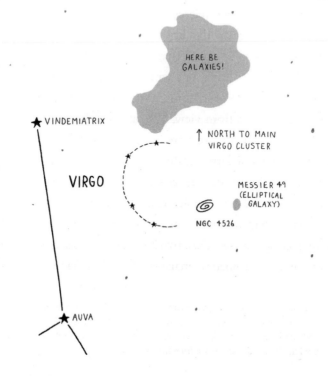

goddess holds at her hip, and forms the southeastern corner of a wonky rectangle making up Virgo's body (see page 302 for a chart). The bulk of Virgo's galaxies lie to the northwest of this shape, straddling the border with the neighbouring constellation of Coma Berenices*, and are well worth a look with good binoculars or a telescope. To find NGC 4526, however, first look for magnitude 3.4 Auva on the opposite side of the rectangle to Spica. From here, you can track to the wonderfully named Vindemiatrix, a brighter yellow-orange giant of magnitude 2.9 just to its north. Scan the space to the east of the Auva-Vinedmiatrix line and you should find a small semi-circle of stars on the edge of naked-eye visibility. NGC 4526 sits at what would be the four o'clock, southwestern edge if the circle were completed.

Through amateur telescopes, 4526 looks like an elongated cigar of fuzzy light with a bright centre spot. Larger instruments and long exposures reveal that this is just the core region of a large lenticular galaxy – one with a central hub and surrounding stellar disc, but no spiral arms, and a thick loop of dust blocking most of the disc from view. When SN 1994D was at its peak magnitude of 10.7, it appeared as a brilliant flare on the galaxy's northwestern outskirts, outshining all of its other stars combined. Before we come to the supernova, however, let's linger a while in the Virgo cluster as a whole, which has secrets of its own.

Galaxy clusters come in all shapes and sizes – our own Local Group gang consists of just two bullying spiral galaxies (the Milky Way and Andromeda), makeweight spiral Messier 33, and a

---

* Queen Berenice's Hair – a rare example of a constellation named after a historical figure. Berenice was a Libyan queen and wife of Egyptian ruler Ptolemy III, said around 245 BCE to have offered her locks of hair as a sacrifice to the gods if they ensured her husband's safe return from his latest war.

bunch of small hangers-on. The Virgo cluster, on the other hand, is a far more formidable proposition, containing dozens of big, bright systems with an entire armada of dwarf galaxies in support. Look out especially for Messier 87, a vast "giant elliptical" ball of stars that squats at the cluster's centre and forms its gravitational anchor – you'll find it shining at magnitude 8.8 about ten Moon-widths north of NGC 4526 (and it's worth a try with binoculars on dark, moonless skies).

As an outlying galaxy, NGC 4526 is trapped on a long, slow orbit around the cluster's centre of gravity, which coincides fairly precisely with Messier 87. But its motion isn't *quite* what you'd expect, and thereby hangs a tale.

Remember Fritz Zwicky, the Swiss astrophysicist who, with Walter Baade in 1935, proposed the existence of supernovae as a separate class of exploding star? Well a couple of years before that, Zwicky made another important discovery – albeit one that would be largely neglected through his lifetime*. This was just a few years after Hubble had confirmed the existence of independent galaxies beyond the Milky Way and revealed the relationship between their distance and speed of retreat, so astronomers were still getting to grips with how to interpret the countless "spiral nebulae" strewn across the sky.

In 1933, Zwicky set out to review and expand Hubble's work on the red shifts in light from distant galaxies caused by the expansion of the Universe. Hubble's own attempts to refine the relationship had foundered somewhat on the fact that the motion of any individual galaxy will not be a pure reflection of cosmic

---

* An infamous grump, Zwicky's favourite insult was to call those who incurred his wrath "spherical bastards" – because they were bastards whatever way you looked at them. You're welcome.

expansion – it will inevitably incorporate local motions due to the gravity of its neighbours and the cluster it sits in. Zwicky realized you could get around this and improve your measurements by averaging out all the different red shifts in a particular cluster.[2]

The technique was a resounding success, and it also revealed, almost as a side effect, the motion of each individual galaxy relative to the average. Zwicky realized this meant you could weigh a galaxy cluster by working out the speed at which its outlying galaxies were orbiting around the centre and using a clever bit of maths called the Virial theorem. When he tried to do this for the bright Coma Cluster, however, the warning klaxons went off immediately – everything was moving much faster than you'd expect based on the amount of visible matter in the cluster, implying that Coma contained about 400 times more mass than its visible stars could account for. Zwicky attributed the missing mass to unseen *dunkle Materie* – dark matter.

At the time, Zwicky's discovery was largely overlooked – estimating the masses of distant galaxies purely from their overall brightness seemed like a bit of a mug's game with so many unknown factors at work. And in the following decades, the amount of missing mass to be accounted for was indeed sharply eroded, as new types of telescope showed that clusters and their individual galaxies contain plenty of matter that's invisible at optical wavelengths, but otherwise perfectly normal, radiating at every wavelength from radio to X-rays.

It wasn't until the 1970s that the problem reared its head once again in an undeniable way. Vera Rubin's measurements of galaxy rotation (touched on briefly in our visit to the star S2), showed

---

* Since on average half of the local motion will be towards Earth (reducing the cosmological redshift) and half will be away from Earth (increasing it).

that the apparently empty outer reaches of the Milky Way and other galaxies wield an unexpected gravitational influence over the orbits of individual stars.

Evidence for dark matter on our galactic doorstep meant that many of the get-out clauses used to explain away the behaviour of distant galaxy clusters would no longer fly. Zwicky's discovery was real, and even if figures for the cosmic missing mass have been whittled down to a slightly less amount, that still means that you, me and all the stars in the Universe are outweighed five to one by dark matter.

So what *is* it? At the most basic level, dark matter is stuff that has mass and makes itself felt through gravity, but which doesn't seem to be detectable in any other way. It's not just dark, and it's not just transparent – it's completely immune to any sort of interaction with light and the other electromagnetic radiations that fill the Universe. We can figure out *where* it is (at least on the largest scales) by measuring its gravitational influence, and the answer seems to be that it's all around us – hanging around individual galaxies and galaxy clusters where normal matter also concentrates. This means that it is probably passing straight through our most sensitive particle detectors (and our own bodies) all the time with nary a hint of its presence.

To save the blushes of physicists, it would be nice if it turned out to just be some compact, dark form of normal matter (stray planets, cooled-down white dwarfs – even black holes would do at a pinch). But evidence suggests otherwise, since numerous experiments aimed at detecting such objects in the halo around our galaxy have come up negative. We're left with the well-intentioned but ultimately somewhat meaningless acronym WIMPs (Weakly Interactive Massive Particles – which describes what they are, but absolutely nothing more), and the hope that some future experiment will put us out of our misery.

Now, about that supernova...

★ ★ ★

When we last discussed these brilliant cosmic outbursts on our visits to the doomed Eta Carinae and the ghostly Crab Pulsar, we made the same natural assumption as Zwicky and Baade – that supernovae are all a result of exploding stars. However, in reality, there are several different types of supernova, most easily distinguished by differences in the rate at which their energy output rises to a peak and then fades away. Most of these variants have now been explained with minimal remixes of the same basic "exploding massive star" theme, but one group of troublemakers requires a completely new approach.

Type Ia supernovae rise more quickly to the peak of their energy output, decline more rapidly afterwards, and are on average also brighter than most other stellar explosions. The spectra of their light shows unique features, and most strangely, they occur in the "wrong" places – often popping up in ball-shaped elliptical galaxies. Such star systems are dominated by low-mass red and yellow stars, which should not have anything like the eight solar masses of material required for a "traditional" supernova*.

In the early 1970s, Brit John Whelan and US astronomer Icko Iben put together a theory to explain what was going on with Type Ia supernovae, based on a combination of precise observations and some clever number-crunching [3]. If you recall our discussion of

---

* Despite being a paid-up member of this squad of cosmic troublemakers, Supernova 1994D's lens-shaped home galaxy NGC 4526 is actually a more plausible location for such an explosion.

nova systems (see RS Ophiuchi for a refresher), they involve gas being transferred from an ageing, bloated star onto the surface of a super-dense white dwarf. The dwarf's newfound "atmosphere" of hot gases eventually becomes so hot and dense that it explodes in a sudden burst of fusion energy.

However, white dwarfs have an upper mass threshold – the famous Chandrasekhar limit of about 1.4 solar masses. Whelan and Iben suggested that if this process of gas transfer was enough to tip a particularly obese white dwarf over that limit, it would destroy itself in an unusual type of stellar explosion. An exploding white dwarf, rich in carbon and oxygen but lacking the heavy elements formed inside heavyweight stars, could explain the unusual Type Ia spectra, but just as importantly it offered a way for two relatively low-mass stars, working in conjunction, to produce a supernova.

What exactly happens during a Type Ia explosion? The most obvious (and intuitively appealing) answer would be that as the white dwarf's gravity finally overwhelms the "degeneracy pressure" between electrons that has been holding it up, the stellar remnant collapses suddenly from about the size of Earth down to a city-scale neutron star, with an accompanying outpouring of energy and particles.

So far, however, astronomers have completely failed to find neutron stars left behind by Type Ia events – and this suggests a somewhat messier cause. White dwarfs are the end state of more-or-less Sun-like stars that cannot continue their nuclear fusion reactions beyond burning helium, which is why they end up dominated by carbon and oxygen (the products of that fusion reaction). But as the star gets very close to the Chandrasekhar limit, conditions inside may become so extreme that this is no longer true. Computer modelling suggests that once a certain critical threshold is crossed, the fusion of carbon into heavier elements finally becomes viable. But because the white dwarf's matter is in a strange "degenerate" state (see Sirius B for more), it doesn't simply expand to accommodate the sudden new source of energy. Instead, the outer layers continue to press down, keeping the lid on a pressure cooker even as fusion rips through the star's interior – until eventually the entire star simply blows itself to smithereens.

* * *

But what does an exploding star, however unusual, have to say about the fate of the entire Universe? Well in the 1980s, as they figured out the details of the Type Ia explosion, astrophysicists realised it offered a unique tool to investigate the most distant parts of the cosmos. Peaking at about 5 billion times the brightness of the Sun, these exploding stars are so bright that they can be spotted in the most distant galaxies, and their light curves allow them to be easily distinguished from other types of supernova.

Most importantly, though, Type Ia supernovae are, at least in theory, always the same: they involve the complete obliteration of 1.4 solar masses of material, and whatever the details of the process involved, it's not unreasonable to assume that exactly the same amount of energy is released in every case. This makes them ideal "standard candles" – the term astronomers love to use for any object with a known or predictable brightness that allows them to peg a distance scale on the Universe.

The best-known standard candles, of course, are the Cepheid variables that Hubble used to discover the scale of the extragalactic Universe (and make a first stab at the rate of cosmic expansion) in the 1920s. By the 1990s, improvements to both hardware and theory had vastly improved both our understanding of Cepheid behaviour and our ability to spot them in more and more distant galaxies, but they are still just rather bright stars, so there are limits to how far away they can be seen[*].

Type Ia supernovae are visible over much greater distances and are much easier to calibrate than Cepheids (while they don't all peak at exactly the same luminosity, there's a very simple formula to work out what level they peaked at). So it's no wonder that two teams independently had the idea of using them to double-check the scale and expansion of the Universe.

The Supernova Cosmology Project (led by Saul Perlmutter at Lawrence Berkeley National Laboratory) and the High-Z[†] Supernova Search Team (founded by Australian Harvard researcher Brian Schmidt, Nicholas Suntzeff of Chile's Cerro Tololo Inter-American Observatory, and soon joined by Adam Riess of UC Berkeley) both set out to track down supernovae in distant galaxies in the mid-1990s. The principle behind the projects was simple: find a Type Ia supernova in a distant galaxy (SN 1994D, alas, was too nearby to be useful); figure out its peak brightness and use that to calculate its distance from Earth; then compare that with the distance suggested by its red shift and the "Hubble Constant" – the standard measure of cosmic expansion based on Cepheids.

If they were lucky, the researchers thought, they might not

---

[*] The Hubble Space Telescope spent much of its first decade in orbit slogging away at its "Key Project" of measuring 30,000 Cepheids in galaxies out to about 70 million light years, in order to get the most accurate measurement yet of the cosmic expansion rate and, by extension, the age of the Universe.

[†] Z in this case is astronomical shorthand for redshift.

only provide a useful "belt and braces" confirmation of the Hubble Constant in the local Universe, but also find evidence for the slowing-down of the Universe since the Big Bang. This would involve the most distant supernovae appearing brighter than expected – the red shift (measured in today's terms) would suggest a certain distance, but because cosmic expansion has slowed over time the supernova would actually be a little closer and brighter.

A measure of the rate at which the Big Bang was unwinding would help to resolve the long-running debate over the balance between cosmic expansion and the total mass of the Universe (both its visible and its dark matter). If the Universe had enough mass, then gravity might eventually overcome expansion and pull everything back to a vast implosion (usually called a Big Crunch*), but if it didn't, then the Universe would expand forever, slowly running down over trillions of years in a long, dark Big Chill.

It's safe to say that neither team expected to find what they did: the most distant supernovae were consistently *fainter* than their red shifts predicted.

After years of cross-checking, building their evidence with more supernovae, and realising they had found the same effect, the teams went public in 1998[4][5]. The Universe, it seems, is not slowing down – it's speeding up.

Wait. What?

The expansion of the Universe is not remaining constant as space stretches apart, still driven by initial impetus from the Big Bang. Neither is it slowing down, as the gravity of visible and dark matter begins to pull things towards each other and overwhelm

---

* I prefer Zaphod Beeblebrox's description of the end of the Universe as "a Gnab Gib" from Douglas Adams' *The Restaurant at the End of the Universe*

that initial Big Bang blast. Nope, it's speeding up – *something* is pushing the boundaries of the Universe away from us at an ever-accelerating rate.

Shortly after the discovery, cosmologist Michael Turner borrowed Zwicky's phrasing and coined the name "dark energy" to describe this something. The term's become ubiquitous, but it doesn't really say much about what it is (any more than, say, "dark matter" or "WIMPs"). However, most ideas about dark energy follow one of two approaches.

The first, broadly known as the quintessence[*] school of thought, sees dark energy as a "scalar field" (physics-speak for something with strength but no direction, that can vary from place to place in the Universe)[6]. In this model, dark energy could be something to do with either visible or dark matter – perhaps an "anti-gravity" fifth force[†] that pushes things unexpectedly apart on the largest scales, or something else entirely.

The second, more single-minded, approach is a revival of Einstein's "cosmological constant"[‡]. In this approach, dark energy comes fitted as standard to the spacetime that makes up the fabric of the Universe – add more space by growing the Universe, and you inevitably get more dark energy. This seems to match up pretty well with reality – more refined measurements since the millennium have suggested that the expansion of the post-Big Bang Universe

---

[*]   Literally *Fifth Element* – though its originators probably had medieval alchemy, rather than Luc Besson's 1997 sci-fi campfest, on their mind. Probably.

[†]   The four known fundamental forces of the Universe, for the record, are gravitation, electromagnetism and the weak and strong nuclear forces.

[‡]   A force or expansion that would prevent spacetime from collapsing inwards due to the mass contained within the Universe, allowing the cosmos to survive forever (in line with the general scientific thought of the time).

did originally slow down as expected, but dark energy began to push back 9 billion years ago, and acceleration took hold in earnest 6 billion years ago.

If the cosmological constant approach is right, and dark energy really is getting stronger over time, then will it eventually fade again? Well, we can't be sure, but if it doesn't then the Universe is doomed to a different fate – not a Crunch or a Chill, but a dramatic Big Rip. At present, dark energy only makes itself known across many billions of light years, but as space expands and the dark energy effect grows ever larger, it will eventually start to weaken the forces holding the Universe together. Galaxy clusters will be the first to feel it, gradually losing their grip on outlying members such as NGC 4526, before disintegrating completely. Then individual galaxies will fall apart as their internal gravity is

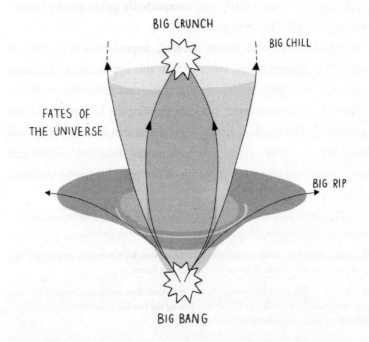

BIG CRUNCH

BIG CHILL

FATES OF
THE UNIVERSE

BIG RIP

BIG BANG

weakened, sending stars drifting off on lonely journeys through intergalactic space. As the process accelerates, solar systems will come loose, and even the gravity that holds individual stars and planets together will give way, before matter itself is torn to shreds on the subatomic level leaving nothing else behind. Fortunately, such events, if they happen, are a long way off – our species, and indeed our entire solar system, won't be around to worry about it.

We don't yet know how things will play out – as we've seen throughout our explorations of these 21 stars (and three imposters), our understanding of the Universe sometimes crawls at a snail's pace, and at other times advances in leaps and bounds. But 20 years isn't a long time to accommodate something as fundamental as dark energy (we've been confident of dark matter's existence for almost 50 and are still mostly advancing by ruling things out).

And even in the past few months, the story has taken another turn, with a team of South Korean astronomers suggesting in December 2019 that perhaps dark energy isn't all its cracked up to be. Young-Wook Lee's group say they've found evidence for Type Ia supernovae growing brighter over the history of the Universe. If they're right, then the faintness of the more distant ones (with the longest lookback time) could perhaps be explained without resorting to dark energy[7]. Could the discovery, which won a Nobel Prize in Physics for Pelmutter, Schmidt and Riess, turn out to be built on sand after all?

It's too soon to say. Since 1998, cosmologists have found independent signs of accelerating cosmic expansion in a number of odd corners of the Universe. The unexpected faintness of Type 1a supernovae remains the central pillar of evidence for dark energy, but if it does fall, it's going to leave a lot of awkward questions in its wake.

So far this is just one paper whose findings need to be verified and explained – and in early 2020 the world has had other things on its mind. Throughout this book we've seen on many occasions how intelligent questions and lateral thinking can shift our perspective on the nature of individual stars, and ultimately on the Universe as a whole – will this be another example, or one of the countless false alarms left forgotten in the race to establish a consensus view?

One thing, however, can be said. If the dark energy hypothesis is successfully challenged, then all bets regarding the fate of the Universe are off. The scientific Tarot deck that reveals our cosmic fate will once again be flung into the air, and no one can tell how its cards will land.

Despite this, we can be sure that the stars will tell us the answers if we only know how to read them. It will be the job of future stargazers to understand their language.

CHAPTER NOTES

1 – POLARIS

1 *Magnitudes of Thirty-six of the Minor Planets for the First Day of each Month of the Year 1857*. Pogson, 1856

2 *Direct Detection of the Close Companion of Polaris with the Hubble Space Telescope.* Evans *et al.*, 2008

3 *Toward Ending the Polaris Parallax Debate: A Precise Distance to Our Nearest Cepheid from Gaia DR2*. Engle, Guinan & Harmanec, 2018

4 *Polaris: Amplitude, Period Change, and Companions*. Evans, Sasselov & Short, 2002.

2 – 61 CYGNI

1 *The Parallax of 61 Cygni*. Hopkins, 1916

2 *Measurement of the distances of the stars*. Dyson, 1915

3 *A letter… giving an account of a newly discovered motion of the fix'd stars.* Bradley, 1728

4 *On the Parallax of Sirius*. Henderson, 1840

5 *A letter from Professor Bessel to Sir J. Herschel, Bart*. Bessel, 1838

3 – ALDEBARAN

1 *On the Spectra of Some of the Fixed Stars*. Huggins & Miller, 1864

2 *Fr. Secchi and stellar spectra*. McCarthy, 1950

3 *Memoir of Henry Draper*. Barker, 1888

4 – MIZAR (AND FRIENDS)

1 *Catalogue of 500 new Nebulae, nebulous Stars, planetary Nebulae, and Clusters of Stars; with Remarks on the Construction of the Heavens.* Herschel, 1802

2 *Continuation of an Account of the Changes That Have Happened in the Relative Situation of Double Stars.* Herschel, 1804

3 *Preliminary Paper on Certain Drifting Motions of the Stars.* Proctor, 1869

4 *On the Spectrum of Zeta Ursae Majoris.* Pickering, 1890

5 *The spectroscopic binary Mizar.* Vogel, 1901

6 *Discovery of a faint companion to Alcor using MMT/AO 5 μm imaging.* Mamajek *et al.*, 2010

5 – ALCYONE AND HER SISTERS

1 *On the use of photographic effective wavelength to determine colour equivalence.* Hertzsprung, 1911

2 *On a relationship between brightness and spectral type in the Pleiades.* Rosenberg, 1910

3 *The Parallax of the Hyades, derived from Photographic Plates.* Kapteyn and De Sitter, 1909

4 *Relations Between the Spectra and Other Characteristics of the Stars.* Russell 1914

5 *The distance of the Pleiades.* Pickering, 1918

6 *The Pleiades (George Darwin Lecture).* Hertzsprung, 1929

6 – THE SUN

1 *New investigations regarding the period of sunspots and its significance.* Wolf, 1852

2 *Stellar Atmospheres; a Contribution to the Observational Study of High Temperature in the Reversing Layers of Stars.* Payne, 1925

3 *On the Composition of the Sun's Atmosphere.* Russell, 1929

4 *Energy Production in Stars.* Bethe, 1939

7 – THE TRAPEZIUM AND OTHER WONDERS

1 *Emanuel Swedenborg – An Eighteenth century cosmologist.* Baker, 1983

2 *Astronomical Observations Relating to the Sidereal Part of the Heavens, and Its Connection with the Nebulous Part; Arranged for the Purpose of a Critical Examination.* Herschel, 1814

3 *On the Spectrum of the Great Nebula in the Sword-Handle of Orion.* Huggins, 1865

4 *On the spectrum of the nebula in the Pleiades.* Slipher, 1912

5 *i Orionis—Evidence for a Capture Origin Binary.* Bagnuolo *et al.*, 2001

8: T Tauri

1 *Extract from a Letter of Herr Hind to the Editors.* Hind, 1852
2 *Note on Hind's Variable Nebula in Taurus.* Burnham, 1890
3 *On the variable nebulæ of Hind and Struve in Taurus, and on the nebulous condition of the variable star T Tauri.* Barnard, 1895
4 *Discovery of an infrared companion to T Tau.* Dyck, Simon & Zuckerman, 1982
5 *T Tauri Variable Stars.* Joy, 1945
6 *Small Dark Nebulae.* Bok & Reilly, 1947
7 *Star Formation in Small Globules: Bart BOK Was Correct!* Yun & Clemens, 1990
8 *Embryonic Stars Emerge from Instellar "EGGS".* Hester & Scowen, 1995

9: Proxima Centauri

1 *A Faint Star of Large Proper Motion.* Innes, 1915
2 *A small star with large proper motion.* Barnard, 1916
3 *On the relation between mass and absolute brightness of components of double stars.* Hertzsprung, 1923
4 *A Third Flare of L726-8B.* Luyten , 1949
5 *Proxima Centauri as a Flare Star.* Shapley, 1951
6 *Observations of the Faint Dwarf Star L 726-8.* Joy & Humason, 1949
7 *A terrestrial planet candidate in a temperate orbit around Proxima Centauri.* Anglada-Escudé *et al.*, 2016
8 The First Naked-eye Superflare Detected from Proxima Centauri. Howard *et al.*, 2018

10: Helvetios

1 *Astrometric study of Barnard's star from plates taken with the 24-inch Sproul refractor.* Van de Kamp, 1963
2 *A planetary system around the millisecond pulsar PSR1257 + 12.* Wolszczan & Frail, 1992
3 *A Jupiter-mass companion to a solar-type star.* Mayor & Queloz, 1995

11: Algol

1 *A Series of Observations on, and a Discovery of, the Period of the Variation of the Light of the Bright Star in the Head of Medusa, Called Algol.* Goodricke, 1783
2 *Dimensions of the Fixed Stars, with Special Reference to Variables of the Algol Type.* Pickering, 1880
3 *Spectrographic Observations of Algol.* Vogel, 1890
4 *A Spectrophotometric Study of Algol.* Hall, 1939
5 *Detection of the Secondary of Algol.* Tomkin & Lambert, 1978
6 *On the Subgiant Components of Eclipsing Binary Systems.* Crawford, 1955
7 *Stellar Encounters with the Oort Cloud Based on Hipparcos Data.* Sánchez *et al.*, 1999

12: MIRA

1 *History of the Discovery of Mira Stars.* Hoffleit, 1997

2 *Johannes Hevelius (1611-1687) – Astronomer of Polish Kings.* Szanser, 1976

3 *Interferometric observations of the Mira star o Ceti with the VLTI/VINCI instrument in the near-infrared.* Woodruff *et al.*, 2004

4 *The color-magnitude diagram of the globular cluster M 92.* Arp, Baum & Sandage, 1953

5 *A turbulent wake as a tracer of 30,000 years of Mira's mass loss history.* Martin *et al.*, 2007

6 *Omicron Ceti a Visual Binary.* Aitken, 1923

13: SIRIUS AND ITS SIBLING

1 *On the Parallax of Sirius.* Henderson, 1840

2 *On the Variations of the Proper Motions of Procyon and Sirius.* Bessel, 1844

3 *The Spectrum of the Companion of Sirius.* Adams, 1915

4 *On the Relation between the Masses and Luminosities of the Stars.* Eddington, 1924

5 *On Dense Matter.* Fowler, 1926

14: RS OPHIUCHI

1 *On the New Star in Auriga.* Huggins, 1892.

2 *A Probable New Star, RS Ophiuchi.* Cannon & Pickering, 1905.

3 *A Photometric Investigation of the Short-Period Eclipsing Binary, Nova DQ Herculis (1934).* Walker, 1956

4 *Binary Stars among Cataclysmic Variables.* Kraft, 1962–64.

5 *Spectroscopic orbits and variations of RS Ophiuchi.* Brandi *et al.*, 2009.

6 *On the Cause of the Nova Outburst.* Starrfield, 1971

7 *The Galactic Nova Rate Revisited.* Shafter, 2017

15: BETELGEUSE

1 *The Parallax of α Orionis.* Schlesinger, 1921

2 *The Internal Constitution of the Stars.* Eddington, 1920

3 *Measurement of the Diameter of α Orionis with the Interferometer.* Michelson & Pease, 1921

4 *Interferometric observations of the supergiant stars alpha Orionis and alpha Herculis with FLUOR at IOTA.* Perrin *et al.*, 2004

5 *Limitations à la Qualité des Images d'un Grand Télescope.* Texereau, 1963

6 *Attainment of Diffraction Limited Resolution in Large Telescopes by Fourier Analysing Speckle Patterns in Star Images.* Labeyrie, 1970

7 *Detection of a bright feature on the surface of Betelgeuse.* Buscher *et al.*, 1990

8 *On the Variability and Periodic Nature of the Star α Orionis.* Herschel, 1840

9 *The Changing Face of Betelgeuse.* Wilson, Dhillion & Haniff, 1997

16: Eta Carinae

1 *The historical record of η Carinae I. The visual light curve, 1595–2000.* Frew, 2004
2 *Eta Carinae.* Gaviola, 1949 & 1952
3 *Light echoes reveal an unexpectedly cool η Carinae during its nineteenth-century Great Eruption.* Rest *et al.*, 2012
4 *On Nuclear Reactions Occurring in Very Hot Stars.* Hoyle, 1954
5 *The 5.52 Year Cycle of Eta Carinae.* Damineli, 1996

17: The Crab Pulsar

1 *Early drawings of Messier 1: pineapple or crab?* Dewhirst, 1983
2 *Observed Changes in the Structure of the "Crab" Nebula (N.G.C. 1952).* Lampland, 1921
3 *Novae or Temporary Stars.* Hubble, 1921
4 *On Super-Novae.* Baade & Zwicky, 1934
5 *The Crab Nebula.* Baade, 1942
6 *Observation of a Rapidly Pulsating Radio Source.* Hewish, Bell *et al.*, 1968
7 *Energy Emission from a Neutron Star.* Pacini, 1967

18: Cygnus X-1

1 *Cosmic X-ray Sources.* Friedman, 1969
2 *Identification of Cygnus X-1 with HDE 226868.* Bolton, 1972
3 *Cygnus X-1—a Spectroscopic Binary with a Heavy Companion?* Webster & Murdin, 1972
4 *On the Means of Discovering the Distance, Magnitude, &c. of the Fixed Stars...* Michell, 1774
5 *On the Gravitational Field of a Point Mass under the Einstein Theory.* Schwarzschild, 1916
6 *On Massive Neutron Cores.* Oppenheimer & Volkoff, 1939
7 *The Rotation of Cosmic Gas Masses.* Weizsäcker, 1948
8 *The Extreme Spin of the Black Hole in Cygnus X-1.* Gou *et al.*, 2011

19: Eta Aquilae

1 *A series of observations on, and a discovery of, the period of the variation of the light of the star marked δ by Bayer, near the head of Cepheus.* Goodricke, 1786
2 *Observations of a New Variable Star.* Pigott, 1784
3 *1777 Variables in the Magellanic Clouds.* Leavitt, 1908
4 *Periods of 25 Variable Stars in the Small Magellanic Cloud.* Pickering, 1912
5 *On the Spatial Distribution of Variables of the Delta Cephei Type.* Hertzsprung, 1913
6 *On the Nature and Cause of Cepheid Variation.* Shapley, 1914
7 *The Internal Constitution of the Stars.* Eddington, 1926
8 *Physical Basis of the Pulsation Theory of Variable Stars.* Zhevakin, 1963

20: IMPOSTER #1: OMEGA CENTAURI

1 *An Account of several Nebulae or lucid Spots like Clouds, lately discovered among the Fixt Stars by help of the Telescope.* Halley, 1716
2 *A Discussion of Variable Stars in the Cluster w Centauri.* Bailey, 1902
3 *Studies based on the Colours and Magnitudes in Stellar Clusters. XII: Remarks on the Arrangement of the Sidereal Universe.* Shapley, 1919
4 *The color-magnitude diagram for the globular cluster M3.* Sandage, 1953

21: S2

1 *An Original Theory or New Hypothesis of the Universe.* Wright, 1750
2 *Catalogue of a second thousand of new nebulæ and clusters of stars; with a few introductory remarks on the construction of the heavens.* Herschel, 1789
3 *The Discovery of Sgr A*.* Goss, Brown & Lo, 2003
4 *Apparent Proper Motion of the Galactic Center Compact Radio Source and PSR 1929+10.* Backer & Sramek, 1982
5 *High Proper Motion Stars in the Vicinity of Sgr A*: Evidence for a Supermassive Black Hole at the Center of Our Galaxy.* Ghez et al., 1998

22: IMPOSTER #2: THE ANDROMEDA NEBULA

1 *Abd al-Rahman al-Sufi and his book of the fixed stars: a journey of re-discovery.* Ihsan, 2012
2 *Novae in Spiral Nebulae and the Island Universe Theory.* Curtis, 1917
3 *Radial Velocity Observations of Spiral Nebulae.* Slipher, 1917
4 *The Scale of the Universe.* Shapley & Curtis, 1921
5 *Cepheids in Spiral Nebulae.* Hubble, 1925
6 *The Resolution of Messier 32, NGC 205, and the Central Region of the Andromeda Nebula.* Baade, 1944

23: IMPOSTER #3: 3C 273

1 *3C 273: A Star-Like Object with Large Red-Shift.* Schmidt, 1963
2 *A Relation between Distance and Radial Velocity among Extra-Galactic Nebulae.* Hubble, 1929
3 *Discussion on the Evolution of the Universe.* Lemaître, 1927
4 *Galactic Nuclei as Collapsed Old Quasars.* Lynden-Bell, 1969
5 *The Beginning of the World from the Point of View of Quantum Theory.* Lemaître, 1931

24: SUPERNOVA 1994D

1 *Supernova 1994D in NGC 4526.* Treffers *et al.*, 1994

2 *Red shifts of extragalactic nebulae.* Zwicky, 1933

3 *Binaries and Supernovae of Type I.* Whelan & Iben, 1973

4 *Measuring Cosmological Parameters with High Redshift Supernovae.* Perlmutter, 1998

5 *An Accelerating Universe and Other Cosmological Implications from SNe IA.* Riess, 1998

6 *Cosmological Imprint of an Energy Component with General Equation of State.*
Caldwell, Dave & Steinhardt, 1998

7 *Early-type Host Galaxies of Type Ia Supernovae. II. Evidence for Luminosity Evolution in Supernova Cosmology.* Yijung Kang *et al.*, 2019

Acknowledgements

*This book came together in strange times, and would not have been possible without the help of a whole bunch of people operating out of home offices and makeshift living-room set-ups in somewhat stressful circumstances. Thanks to the entire crew at Welbeck Publishing for sticking with it!

A special nod at Welbeck to Wayne Davies – for remembering me from a previous life and getting it off the ground, and to the unflappable Oli Holden-Rea for keeping a firm hand on the tiller and steering me through the whole experience.

Elsewhere, huge thanks to Nathan Joyce for his thorough but sympathetic editing, to the fantastic Laura Barnes for transforming my somewhat ropy sketches and notes into illustrations that look so wonderful on the page, and to Graeme Andrew at Envy Design for bringing the whole thing together.

Finally, thanks to all the friends and family who kept me on the straight and narrow with their requests for constant progress reports – I hope you all enjoy the end result.*